DEEP SKY
OBJECTS

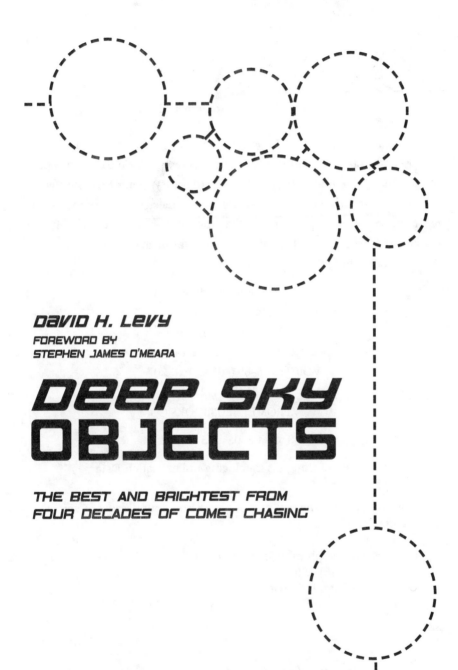

DAVID H. LEVY

FOREWORD BY
STEPHEN JAMES O'MEARA

DEEP SKY
OBJECTS

THE BEST AND BRIGHTEST FROM
FOUR DECADES OF COMET CHASING

PB Prometheus Books

59 John Glenn Drive
Amherst, New York 14228-2197

Published 2005 by Prometheus Books

Inquiries should be addressed to
Prometheus Books
59 John Glenn Drive
Amherst, New York 14228–2197
VOICE: 716–691–0133, ext. 207
FAX: 716–564–2711
WWW.PROMETHEUSBOOKS.COM

09 08 07 06 05   5 4 3 2 1

Library of Congress Cataloging-in-Publication Data

Levy, David H., 1948–
    Deep sky objects : the best and brightest from four decades of comet chasing / by David H. Levy.
        p. cm.
    Includes bibliographical references and index.
    ISBN 1–59102–361–0 (pbk.: alk. paper)
    1. Comets—Observer's manuals. I. Title.

QB64.L478 2005
523.8—dc22

2005020473

Printed in the United States of America on acid-free paper

For Wendee—
Thank you for making our lives so special.
I love you.

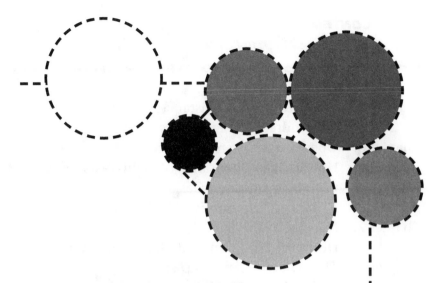

CONTENTS

PART 2: THE BEST AND THE BRIGHTEST 47

PART 3: THE FULL LEVY LIST   265

ACKNOWLEDGMENTS

To Tim Hunter, who has made a great effort to photograph some of the objects;

To Dean Koenig, whose photographs have enriched this book;

To Jack Newton, Scott Roberts, and Tom Glinos, who have also contributed photographs;

To Edwin Aguirre, Stephen James O'Meara, and Rick Fienberg at *Sky &Telescope*, for their editorial suggestions, and for *Sky & Telescope*'s permission for me to adapt some of my *Star Trails* columns;

To Bernard Arghiere, Simon Chung, Leo Enright, Lance Humphreys, Peter Jedicke, Sid Leach, Bill Logan, Steve O'Meara, Alex Scheeline, and Robert Summerfield, who assisted with

particular objects, and to my nephew and godson Michael Levy, who assisted with the Web site http://www.jarnac.org/ levylist.htm associated with the catalog.

To my friends at Prometheus Books—this is my third book with them;

To our children Nanette and Mark;

To Wendee, whose role in the last decade toward the completion of this book has been outstanding—she has spent many, many hours helping with the observing, photography, writing, editing, choice of photographs, and lack of sleep as our deadlines approached;

To my readers—to all of you, my sincerest thanks for believing in this project.

Finally, thanks to our grandchildren, Summer and Matthew, who, as shown in the photograph below, have helped with the final editing of the manuscript.

Assisted by his grandchildren, Matthew (left) and Summer, the author works on the final stages of his book.

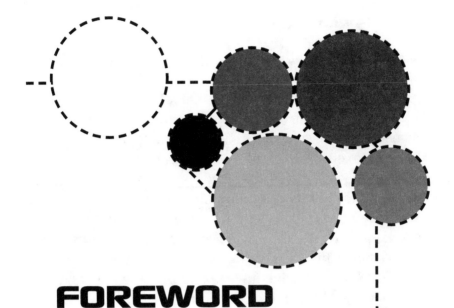

FOREWORD

One clear night in April 1988, my wife, Donna, and I were sitting under the stars on David Levy's roof. David was not with us. An hour earlier, he had climbed down the vertical ladder leading from the roof to the ground, walked a few yards to the south, and locked himself in his Jarnac Observatory—a rectangular wooden structure that housed Miranda, his 16-inch Dobsonian comet seeker.

As the minutes passed, Donna and I fell silent as David, now out of sight, patiently swept the heavens for those cosmic hairballs we call comets. Suddenly, we heard a muffled cry from David: "Oh, hello," he said. Jumping up, because we thought David had another guest,

we scanned the yard from our high perch, but saw no one. All was still. Perhaps David was on the phone. Minutes passed again in silence, until we heard, "Oooo! Wow, You look nice!" Once again, we were up and scanning the yard in vain. Then it dawned on us. David was not entertaining any normal guest. He was conversing with the sky. He was reacting to the sudden appearance of a razor-sharp, edge-on galaxy or a globular cluster, where hundreds of thousands of suns are packed into a tiny globe of scintillating light. These are David's friends, our celestial neighbors. David is the consummate romantic. He has two passions: his wife, Wendee, and the stars above. If you want to see David's soul, just look up into the night sky.

When David told me he was going to write a book about his favorite deep sky objects, a smile came over my face. I cannot think of a better person to share with you the objects he has come to love and adore for the last half century. David's list is not meant to supplant any existing lists. It is an index of celestial artwork. What David has done is akin to opening a museum that displays the works he has visually "collected" over the years. All he wants to do is to share these wonders with you—from his discerning eye.

Astronomy is such a personal endeavor. No two people looking at the sky will fall in love with the same deep sky objects, which is why it's fun to share. David's list represents the biggest and best, especially those objects that have inspired David the most. It is a portrait of a collector. By looking at these objects we can gain a better understanding of what it is that makes us one with the sky. Once you discover why David has selected these objects, perhaps you, too, will be inspired to create your own list and share it with others. As with museum art, the artist does not define the art, the art defines the artist. So go out and enjoy. David will be your guide.

Stephen James O'Meara   Volcano, Hawaii, 2005

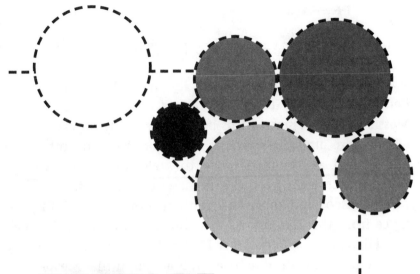

PREFACE

As he who studies fervently the skies
Turns oftener to the stars than to his book . . .
　　　—Lord Byron, *Don Juan* 2.163, 1819

It's the night sky. The Sun has set, the sky is clear, our telescope is ready, and one by one, the stars are beginning to come out. Welcome to the show that never ends.

This is the night sky that Charles Messier observed from a distant time and place. More than two hundred years ago, from Paris, Messier became the first observer to search for and discover comets as part of an organized program; he kept records of the celestial objects that appeared through his telescope during the nights

of his search. After ten years, he published the first version of his catalog.

Why did Messier publish his list in 1771? As translated in Ken Graun's *The Next Step: Finding and Viewing Messier's Objects*,[1] Messier wrote: "I started this book in 1764, by observing those which were already known as well as researching others that had eluded astronomers since the telescope was invented." However, he then proceeds to describe his motivation: "[In 1758] I discovered above the southern horn and at a short distance from the Zeta star of this constellation, a whitish, elongated light. . . . On September 11, 1760, I discovered in the head of Aquarius a beautiful nebula." It is apparent that Messier's first two objects inspired him to build a list, but that over the following decade, the list and its purpose took on a life of its own.

I wish that more comet hunters had followed Messier's lead and had published lists of their own "comet masqueraders" as Leslie Peltier, one of the most famous comet finders of the last century, put it. Although Peltier did keep a list of those objects arranged in order of right ascension, he never published it. It would be fascinating to see which objects were common to all the lists, whether published or not, kept by different searchers. Because comet searchers use different types of telescopes under different types of sky, the number of common objects would probably be relatively low.

My own list has the same inspiration as Messier's, and this book is similar to the stated purpose of Messier's first catalog. But astronomy has changed utterly since 1771, and major catalogs now cover virtually everything in the sky down to about 20 magnitude. While Messier's list was intended to guide astronomers, mine is to guide beginners into the field of deep sky objects, and to offer a personal taste of the night sky to observers. This is a list of objects in the sky that I've been watching for decades, and the objects that this book will feature have come into the eyepiece of my telescope in a procession that began on January 1, 1966, only two weeks after I started my search for comets, with a small cluster called NGC 1931.

It has been almost forty years since that frigid January first. I've spent thousands of hours with my eyepiece, patiently moving from field to field, in my search for comets. That search has been quite successful; over the years I have discovered eight new comets crawling across the sky, as well as thirteen new comets on photographic film. And as much as I have enjoyed the thrill of each of those comet discoveries, I can categorically say that if those moments of personal discovery were the only causes of joy for my program, I would have given it up years ago. It is what I have found on the road to a comet that has kept me going. And it is what I've found on that road that is the subject of this book.

This book is a journey to distant objects in the night sky. Deep sky objects are generally considered to be anything beyond the solar system. But in reality the various lists of deep sky objects that have appeared concentrate on objects so far away that they present a fuzzy appearance in a telescope. The objects I have selected are all over the sky, and they come with interesting historical stories. They include red stars, double and triple stars, unique star patterns, clusters, nebulae, galaxies, and quasars.

Two factors make this book different from other guides to the deep sky. First, it is not a mere listing of objects but an *approach* to the wonders of the deep sky. It includes many objects that can be found even in a light-polluted sky, which means that even someone just starting out in astronomy can enjoy them.

HOW THE LEVY LIST IS BUILT

... none of us so much as know our letters in the stars yet ...
—Charles Dickens, *The Mystery of Edwin Drood*, 1870

The highlights of the list this book offers did not come from hours of reading and research, but from four decades of actually searching the sky. I am a comet hunter. As I would search a region of sky, occasionally I would come across an interesting object—an unusually colored star or cluster of stars, a cloud of gas, or a galaxy. An object has to have something special about it to merit inclusion. Over the years I have located more than three hundred objects, of which the "best and brightest" will be featured in this book. The objects will be presented in the order of their distance from us. This way the reader—particularly the beginning reader—will be taken on a tour of fascinating objects beginning with double and triple stars and the exciting explosive variable stars, then move out toward the open clusters and asterisms, then to the more distant nebulae, then to the galactic center and the globular clusters orbiting our galaxy, and on to galaxies and ultimately to the clusters of galaxies.

Finally, at the end, readers will have access to information on all 378 objects in the list. And once you've put the book down, in the years to come, you can watch as the catalog continues to grow at http://www.jarnac.org. Welcome to my family of deep sky objects. May you enjoy meeting them.

Part I

AN INTRODUCTION
Beyond the Moon

Great father he of generation,
Is rightly cald, th'author of life and light . . .
      —Edmund Spenser, *The Faerie Queene*,
          1596, referring to the Moon

. . . to expect anything better is to ask for the
moon and the stars.
      —Miguel de Cervantes Saavedra,
        Don Quixote de la Mancha, 1605

Thence, to the Circle of the Moone she
clambe . . .
      —Spenser, *Cantos of Mutabilitie*
        (*The Faerie Queene*), 1609

I'm the Knight of the White Moon I am . . .
> —Cervantes, *Don Quixote*

The Moon is down; I have not heard the clock.
And she goes down at twelve.
> —William Shakespeare, *Macbeth* 2.1.2–3, 1606

For, ever since, immortal man hath glowed
With all kinds of mechanics, and full soon
Steam-engines will conduct him to the moon.
> —Lord Byron, *Don Juan* 10.27–8, 1823

He sighed;— the next resource is the full moon, . . .
But Lover, Poet, or Astronomer—
Shepherd, or swain—whoever may behold,
Feel some abstraction when they gaze on her;
Great thoughts we catch from thence . . .
> —Lord Byron, *Don Juan* 16.13.1, 16.14.1–4, 1823

. . . Cloisterham being so beautiful, with the moon shining on it—these things inclined me to open my heart.
> —Charles Dickens, *The Mystery of Edwin Drood*, 1870.

The Moon faded behind a sinister black cloud.
> —Leon Uris, *Exodus*, 1958

The night was misty and there were no stars or moon . . .
> —Uris, *Exodus*

The Moon last night made a porpoise or a dolphin out of black cloud and haloed it with silver.
> —Peggy Pond Church, *The House at Otowi Bridge*, 1959

Even though the Moon is not a deep sky object, it deserves some tribute in this book. Its presence or absence in the night sky dictates the nature of our observing sessions. At two and a half days before and after full phase, its light brightens the sky at half the level that it does at full phase. For those five days, only the brightest deep sky objects, like the Pleiades, are worth viewing. Before or after that time, however, it is possible to see a sizable proportion of the objects in this book. Only those objects that require a dark sky (and are so described in the chapters to come) need to be scheduled around the times when the Moon is not in the sky. Indeed, amateur astronomers pay close attention to the phases of the Moon in planning their sessions under the stars.

Besides helping us plan our observing sessions, the Moon is the one object in the sky that has managed to find its way into literature more often than any other. Through the centuries, it has called to us, beckoned to us. It summoned Shakespeare to create the mood for the night that Macbeth killed Duncan, and it inspired Lord Byron to predict the Apollo Moon flights a century and a half before they actually happened. The Moon is one of my favorite objects in the sky. At countless evening star parties, I have watched its light carrying its inspiration into the eye of a child, and while alone, I have climbed its mountains and crawled into its craters.

Like Macbeth, we wait until the Moon goes down to get the darkest possible sky for our "denizens of the deep"; like Uris, we wait for the Moon to vanish and the deeper sky to open its doors to us.

CHAPTER 1

LETTING THE SKY COME TO YOU

Does it seem incongruous to you that a Middlemarch surgeon should dream of himself as a discoverer? Most of us, indeed, know little of the great originators until they have been lifted up among the constellations and already rule our fates. But that Herschel, for example, who "broke the barriers of the heavens"—did he not once play a provincial church-organ, and give music-lessons to stumbling pianists? Each of those Shining Ones had to walk on the earth among neighbors who perhaps thought much more of his gait and his garments than of anything which was to give him a title to everlasting frame: each of them had his little local personal history sprinkled with small tempta-

tions and sordid cares, which made the retarding friction of
his course towards final companionship with the immortals.
—George Eliot (aka Mary Anne Evans),
Middlemarch, 1872

Astronomical discovery has always had a kind of magic to it.
Tycho Brahe's discovery of the supernova of 1572 and his
observations of the comet of 1577 were very important events
in the history of science. Thanks to those events, humanity
learned that a pattern of fixed stars was not always unchanged
and that comets are not, as Aristotle believed, exhalations of air
in Earth's atmosphere, but instead are real objects farther than
the Moon. In 1609 Galileo began making a series of discoveries
that completely changed humanity's understanding of its place
in the cosmos. The moons of Jupiter, the spots on the Sun, and
the phases of Venus all served as evidence that the Sun, not
Earth, is at the center of the solar system.

More than a century later, two great astronomers made
other discoveries that extended humanity's understanding of
its celestial surroundings. William Herschel, about whom
George Eliot wrote so convincingly, is best known for his discovery of Uranus. Besides looking up at the sky, he also looked
up to his colleague from France, Charles Messier, who by 1789
was well known as a comet finder.

At the time of William Herschel's discovery of Uranus,
Messier was celebrated as the first person to discover comets as
part of a deliberate, planned search program. Having found his
first comet in 1760, he became so well known for his finds that
within a few years Louis XV had dubbed him the comet ferret.
More than two hundred years later, Messier's comets were long
gone, but the galaxies, clusters, and clouds of gas and dust that
he found during his search are still there, still precisely in the
sky where he found them, and still easily visible through small
telescopes.

Messier's career is an example of astronomical serendipity. He wanted to find comets, but it is what Messier found *on the road to those comets* that makes him such a highly respected figure today. And his findings on the road comprise a catalog that is so relevant to this book. Messier's catalog is the first list of objects in distant space. Created to decipher which roadblocks to avoid on the way to a comet, the catalog now stands by itself as a way for any new observer to become familiar with what's out there.

CHARLES MESSIER'S CATALOG

Fuzzy objects that are not comets lurk all over the sky. They are beautiful to watch, but for people who search for comets they can be viewed as an inconvenience; comet discoverer Leslie Peltier called them "comet masqueraders." At the end of 1758, Messier found a fuzzy patch of light near the star Beta Tauri. As he studied it from hour to hour and from night to night, he found that the faint fuzzy object stayed plastered to the sky; even though it looked like a comet, it never moved like a comet.

Thus, in his pursuit of comets, and not to be fooled again, Messier decided to include in his catalog each object that he came across during his search program. The first entry, now called Messier 1 or M1, is more popularly known as the Crab Nebula because it resembles a ghostly version of the sea animal. Besides this supernova remnant, Messier's catalog includes a treasure trove of the Northern Hemisphere sky's distant objects. Some of those Messier discovered; others he merely listed. In 1962 I began my own "Messier hunt" with a single observation of the Pleiades, M45. In the spring of 1967, using a larger telescope—an 8-inch f/7 reflector—I finished my list while observing from my grandfather's cottage at Jarnac Pond, Quebec. The last one I found was Messier 83, a beautiful and complex galaxy in Virgo. The most difficult one for me was M61, which I found at the Adirondack Science Camp on July 16, 1966.

Messier published three versions of his catalog. The first 45 objects appeared in 1774, and by 1781 his list had grown to 103. Besides the roster that Messier created for himself, other objects he recorded but never listed were later included, so that now the Messier catalog totals 110 objects spread over much of the sky.

I was well on my way to identifying all the Messier objects when I started my program of comet hunting on December 17, 1965. I had no idea that the adventure would lead as far as it did, to eight visual discoveries so far and thirteen photographic finds. But what happened on that road—my bumping into each of the objects that form this book—might be as important as the search itself.

Perhaps the most basic lesson is that comet searching provides us with an opportunity to explore what the sky really has to offer. Each of the objects in the catalog that forms this book came to me during my search for comets. Thus, each one has a story to tell: a memory of some special observing session when it first entered my eyepiece, compelled me to pause for a moment, then made me take the time to identify the object and add it to my list. This book features these lodestars on the path to comets. What follows now is a summary of the comet search that led to them.

A BACKGROUND OF MY COMET HUNTING PROGRAM

I have been comet hunting for forty years, and over that time my program has evolved. When it began on the night of December 17, 1965 (thirty-nine years to the day before I write this), I did not list the actual finding of a comet as the program's primary aim. In the program log that night, I wrote instead what my program goals were:

(1) To become familiar with the sky through searching for comets and/or novae.
(2) To discover either a comet or a nova.
(3) To learn as much as possible about comets and/or novae through a research program.[1]

I learned a great deal about comet hunting in the months after that chilly December night. The first breakthrough came the following summer, when under the dark sky of the Adirondack Science Camp I was able to spot a faint, large galaxy called M101. I entered that as L2 on my list. It taught me that a dark sky would allow me to see faint galaxies whose surface brightness was less than the background brightness of my light-polluted sky at home. This meant that my search for comets was likely to be more successful if I could find a dark sky.

Even after accomplishing that end—moving to the dark sky near Tucson, Arizona, in 1979—it was not until November 13, 1984, that I discovered my first comet (Comet Levy Rudenko C/1984 V1) near NGC 6009, a cluster of stars. After 917 hours and 28 minutes, spread out over nineteen years, the second aim of my program was achieved at last.

Early in January 1987 I found my second comet (C/1987 A1) as a faint fuzzy visitor on a chilly—and rainy—Tucson morning. The sky was pretty cloudy—in fact within half an hour of the discovery it was pouring rain! I discovered my third comet (C/1987 T1) only 107 observing hours after the second.

On March 20, 1988, I found a comet (C/1988 F1) in the predawn sky only two weeks after I met Gene and Carolyn Shoemaker for the first time. This new comet, it turned out, was virtually identical in its orbit to a comet the Shoemakers found a month later. This was the first case of a pair of related long-period comets being discovered independently. Some twelve thousand years ago the two comets were one that, for some reason, split apart.[2]

Of the next few comets that I found, the most interesting were the comet of 1990, which became bright enough to be seen with the naked eye during that summer, and a periodic comet in 1991. The latter turned out to be a new periodic comet that returns to the vicinity of the solar system every half century. For some reason it had never been picked up, with one possible exception: In 1499 Chinese and Korean observers observed a comet pass from Hercules through Draco, and the Little and Big Dippers.[3] The orbit of that comet is so similar to that of Periodic Comet Levy that it could be the same comet, although positive identification will probably have to wait until the comet returns around 2041.

COMET HUNTING VIA PHOTOGRAPHY

In the fall of 1989 I began a new kind of comet searching. My 1988 meeting with the Shoemakers led me to become a partner in their comet and asteroid search, which took place for a week each month at the 18-inch telescope at Palomar.

On April 15, 1994, I discovered a comet in the tiny constellation of Equuleus. A few hours earlier, the Japanese comet hunter Takamizawa found the same comet using photography as his detective. Then came the summer of 1994, an unforgettable time during which we watched Comet Shoemaker-Levy 9's (D/1994 F2) spectacular encounter with Jupiter.

As I write these words ten years after codiscovering Comet Shoemaker-Levy 9, I have found nothing since then. Back in the freewheeling nights when I was discovering a new comet almost every year, amateur search programs competed mostly with each other. Even when the photographic surveys like the one that the Shoemakers and I conducted using the 18-inch telescope at Palomar were at their height, it was still possible to discover comets visually in the large areas of sky closer to the Sun.

By 1997 well-funded surveys were out of the gate and starting to automatically discover asteroids and comets. These projects include LINEAR, run out of Lincoln Lab; NEAT from NASA's Jet Propulsion Lab; and Spacewatch and the Catalina Sky Survey from the University of Arizona's Lunar and Planetary Lab. These major surveys, as wonderful as they are, have come to mean that amateur visual comet searches are far less likely to discover comets than they were even a few years ago.

MY PRESENT PROGRAM

I am searching for comets today as carefully and as enthusiastically as I've been doing for most of the last forty years. Back in the summer of 1965, I spent many nights learning the sky above the Adirondack Science Camp. I didn't know it at that time, but I was preparing for my comet search, which began a few months later, by becoming as familiar as possible with as many deep sky objects that could be savored under a dark sky. During that summer I was working with other young people, thirteen-year-olds like David Larach. He had arrived at camp eager to study electronics, but after staying outside with me over several magnificent nights decided to focus his interests on astronomy. He has never lost that love. The effort that I made that summer to motivate other young people like David to love the sky has also helped train me for comet hunting.

That was forty years ago. Today I search visually and with modern electronic cameras called charge-coupled devices (CCDs). In the modern CCD mode, I take three or four images of every field, and over the course of a night, this might add up to several hundred fields. The images are scanned for asteroids and comets. My happiest nights, in fact, are when I observe in two ways at once. As I nudge Miranda, my 16-inch reflector, from field to field over a chilly hour or two, or three, at least two other telescopes I use are busy automatically acquiring hundreds of images of other parts of the night sky. This multiple searching technique intensifies my appreciation of what I'm doing: as field after field of stars pass through my eyes, the occasional whirring of motors tells me that my other telescopes are working as well. I feel as though I'm getting double or triple the bang for my buck.

> Tranquility, peaceful surroundings, the pleasures of the countryside, the serenity of the skies . . . to fill the world with wonder and delight.
>
> —Cervantes, *Don Quixote*

CHAPTER 2

GETTING EXCITED ABOUT THE DEEP SKY

Who wouldst not leave him in his wandering
To seek for treasure in the jeweled skies . . .
> —Edgar Allan Poe,
> "Sonnet to Science," 1831

Last night of all,
When yond same star that's westward from
    the pole
Had made his course that part of heaven
Where it now burns, Marcellus and myself,
The bell then beating one—
> —William Shakespeare,
> *Hamlet* 1.1.35–38, circa 1601

Almost three hundred years after Shakespeare wrote these lines, I crawled out of my warm bed on the night of September 1, 1961, trudged down the stairs, opened the side door of our home, and walked out onto our terrace. High in the eastern sky shone the Pleiades, its six bright stars easily seen without a telescope, without binoculars. It was my first view of what we call a deep sky object.

What exactly is the deep sky? Essentially, the term refers to anything in the sky that lies beyond our solar system. When we look at Jupiter, Saturn, our Sun, or even a comet, we're not exploring the deep sky. But when we look at any of the distant stars, we are observing the deep sky.

Another meaningful memory took place on the night of March 23, 1963. For two hours I sketched the positions of 240 stars in the Milky Way, observed the Beehive star cluster Messier 44, and checked on the bright star Sirius. For me, that was a deep sky observing session.

EVERY STAR IS A SUN
(STARS ARE DISCUSSED IN CHAPTER 4)

How does a star work? As gravity forces its material toward its center, thermal pressure tries to drive it outward, keeping the star in equilibrium. At our Sun's core, some four million tons of hydrogen are fused into helium every second. Although this has been going on for some five billion years, less than 6 percent of the Sun's supply of hydrogen has been converted into helium.

The Sun will retain this equilibrium as long as there is hydrogen in its core to provide fuel for its nuclear fires. Other stars, at different stages in their lives, show us what happens when the hydrogen in their cores run out.

Variable Stars

As the famous amateur astronomer Leslie Peltier once noted, a variable star is not just a star that's there, it's a star that's "happening."[1] Of all the wonders that the stars hold, perhaps the most interesting is that some of them change in brightness. We call these stars variable stars. The American Association of Variable Star Observers (AAVSO) collects observations of these stars by amateur astronomers and makes them available to professional astronomers. Delta Cephei is an example: it changes in brightness over a period of five days. It varies because of a change that takes place within the star; as the star expands in size, it fades, and as it contracts, it brightens. Since the star remains bright throughout its cycle, it can be followed each night with the naked eye or with a pair of binoculars. There are other examples of variable stars, from red giant suns that pulsate slowly over periods of many months, to the explosions of novae and supernovae.

Pleiades-like Star Clusters, or Open Clusters [Discussed in Chapter 5]

Like people, stars are born into families, but stars' families are called "clusters" that are open or galactic. We call them "open" because we can see their individual stars or "galactic" because most of them are within our galaxy, rather than orbiting it. Neither open nor galactic really describe these beautiful collections of stars; the famous telescopist John Dobson particularly objects to the term *open cluster*: "Who opened them?" he asks. "Who has their key?" Whatever we call them, since these Pleiades-like clusters are much closer than the globulars, we see them not as fuzzy spots but as masses of individual stars. If you've seen the Pleiades, you've seen an open cluster.

nebulae: clouds of dust and gas [discussed in chapters 6 and 7]

"My God!," exclaimed William Herschel. "There's a hole in the sky!" More than two centuries ago, the man who discovered the planet Uranus found a totally new kind of object. Herschel and his son John, who both imagined that these objects could be doorways to the infinity of space beyond, had found nebulae—matter in space that hides the light from stars behind it.

In the mid-nineteenth century John Herschel observed the southern stars from his observing site just south of Table Mountain in the South African city of Cape Town to prepare for his General Catalog of Nebulae. (In 2003 I had the privilege of observing with my own telescope at the spot where Sir John Herschel first opened up the southern sky almost two hundred years earlier.) Thus, Herschel's holes in the sky were not portals to the great beyond but clouds of gas and dust called nebulae. But because there are no nearby stars to light them, they are dark. They are the same as bright clouds, except that no stars are near them to cause them to glow.

the galactic center [discussed in chapter 8]

I never really appreciated the majesty of our galaxy's center until I saw it against a black sky at an altitude of 12,500 feet. As I stared upward in disbelief, my first thought was: "I can go home. In this brief minute I have truly seen everything." Over the next few nights, I learned that although the true center of our galaxy is hidden under masses of dark nebulae, it is surrounded by the magnificent spectacle of the Milky Way's center, which under a dark Southern Hemisphere sky is really something to behold.

Our galaxy contains some 400 *billion* suns. It looks like a pinwheel, as several spiral arms uncoil to a distance of over one hundred thousand light-years. Surrounding this galactic disk is a halo that stretches at least the same distance farther out. As far away as the halo is, it still contains the globular clusters, some of which we can see through binoculars from our own backyards. Stretching even farther out, and including the space some small neighboring galaxies occupy, is a thin layer called the corona.

GLOBULAR CLUSTERS
[DISCUSSED IN CHAPTER 9]

About one hundred fifty globular star clusters, each containing tens of thousands of stars, lie scattered throughout the sky. Through a small telescope they look like small fuzzy spots, but larger ones resolve the spots into a large number of stars. Globular clusters have been studied almost since the invention of the telescope—in 1665 Abraham Ihle found a large cluster, now called Messier 22, in Sagittarius. More than a century later, William Herschel, in 1786, suggested that these clusters were large groupings of stars. Most of the globular clusters we see are in the Milky Way's outlying regions. A globular cluster can be a hundred light-years wide. Among the oldest things in the galaxy, the stars of the globular clusters are almost as old as the galaxy itself. Some estimates put them as old as 16 billion years.

GALAXIES
[DISCUSSED IN CHAPTERS 10, 11, 12, AND 13]

Our own galaxy is about the same size as the Andromeda Galaxy. As we move ever farther out into space, we find galaxies stretching out without end. The Universe seems to go on and on, and distances seem to get harder to fathom. The farthest thing that we can see in the night sky is the Andromeda

Galaxy, whose distance has been pegged at two million light-years away. It is impossible to imagine a number that large. Light travels 186,272 miles every second; that is the equivalent of about seven times around the world each second. At that speed, light takes two million years to reach us from the Andromeda Galaxy; it left when our earliest ancestors were walking about the Earth.

As far away as that galaxy is, it is one of the closest. In chapter 13 we will explore a quasar, the active core of a galaxy 8 billion light-years away, its light made visible only because it is amplified by the gravity of a closer galaxy, a galaxy acting as a cosmic repeater station. We've come a long way since the supernova of 1572 focused attention on the distant stars. Now we're ready to explore by ourselves this marvelous realm above our heads called the deep sky.

CHAPTER 3

AN OBSERVING GUIDE TO THE CATALOG

'Twas noontide of summer,
  And midtime of night,
And stars, in their orbits,
  Shone pale, through the light
Of the brighter, cold moon,
  'Mid planets her slaves,
Herself in the Heavens,
  Her beam on the waves.

  I gazed awhile
  On her cold smile;
Too cold—too cold for me—
  There passed, as a shroud,
  A fleecy cloud,
And I turned away to thee,
  Proud Evening Star,

In thy glory afar
And dearer thy beam shall be;
For joy to my heart
Is the proud part
Thou bearest in Heaven at night,
And more I admire
Thy distant fire,
Than that colder, lowly light.

—Edgar Allan Poe, "Evening Star," 1827

To enjoy observing deep sky objects, you should account for the following:

(1) *The sky.* Although it is possible to enjoy the Moon and the planets from the light-polluted skies of large cities, deep sky objects can be more finicky. If you observe from a city, you can still delight in all the double stars, variable stars, and clusters of stars that are featured in this book. Only the fainter globular clusters and the galaxies require a dark sky.

(2) *A good telescope.* Today there are more telescopes on the market than at any other time in history. A visit to an amateur astronomy club, especially to one of the observing sessions that they sponsor, will go a long way toward helping choose the telescope that best suits your needs.

For the first few weeks with your telescope, concentrate on the objects that you can first see without a telescope, particularly the Moon and the bright planets. This way, you will get more accustomed to your telescope. You might want to adjust the alignment of the finder—you got it close to perfection on the first night, but maybe it could use some "tweaking." As you get some more experience, you will want to try fainter objects, things that you need a star chart to find. In such cases, find the object first on a star chart. To make the transformation from dots on a printed page to real stars in the sky is a process that takes getting used to. Do it slowly. Find a small group of bright stars on the chart, then find it in the sky, and proceed star by star, from the map to the sky, until you center on the spot in the sky that contains the object you are looking for.

Using the telescope's finder, center a star that is close to the object, preferably within one degree (or the fingertip from your outstretched hand). Now use the main telescope and the lowest-power eyepiece to move the telescope slowly in the direction of the object you seek. Finding your object might take a few tries, but if it is as close to the star as a fingertip with the naked eye, it should be no more than one or two fields of view away in the telescope.

(3) *Appropriate magnification.* High magnification is not always the thing to aim for, and certainly not on the evening of your first look. Generally the highest power most observers use is 60 magnification per inch of the telescope's mirror diameter; that means that 240 power is the maximum useful power for a 4-inch diameter telescope. The higher the power, the harder it is to find the object you're seeking or to keep it in the field of view. Moreover, high powers are far more sensitive to conditions in the atmosphere above you. We call this atmospheric effect "seeing." A very clear night during which the stars are twinkling visibly usually means that high powers will be completely useless, since the object will jiggle or wave like a flag, making it impossible for you to detect any of its details.

(4) *Averted Vision.* Appreciating the fainter objects described in this book takes a little practice. The first time you look at a galaxy, for instance, you may barely see it against the background of sky. But with more experience it will become easier to see. The galaxy seems easier to spot if you look through, as they say, the corner of your eye. Averted vision is the term for this process. Turn your eyes in some other direction, yet still concentrate on the object. Now you're using the more sensitive "rods" that are not in the center of your eye.

RECORDING YOUR OBSERVATIONS

Keeping an observation diary or log is a valuable way to remember the experiences you have with your telescope. Here, by way of example from my own observing log, is one simple way of recording what you see (a translation appears below it):

1518AN4/July 4, 1966/2145–0435/10/Adirondack Science Camp/Pegasus/Andy Bauman, Steve Ashe/Saturn, Albireo, Mizar. Old Messiers: M31, M32, M13, M4, M92. CN3 2 hours. While comet hunting I came upon M101, a Messier I have not seen before. It took me only ten minutes to check it out and to confirm it. At least now CN3 has done something of technical value for me! CN-1, Aurora-1, V.S. 13 (3 new variables, g Herculis, X Herculis, RR Coronae Borealis.)

Translation:

*1518AN4 is the observing session number in a sequence that began with session 1, which was the partial solar eclipse of October 2, 1959. AN means that the session lasted all night; E would mean a session held in evening hours.

*The session began at 9:45 PM and ended at 4:35 AM.

*10 means that the sky conditions, on a scale of 1–10, were near perfect.

*The session was held at the Adirondack Science Camp, on the Twin Valleys campsite owned by the State University of New York at Plattsburgh.

*The telescope used was Pegasus, my 8-inch f/7 (meaning focal length 56 inches) reflector.

*Two of my closest childhood friends, Steve Ashe and Andy Bauman, were observing with me.

*The objects viewed were Saturn, the double stars Albireo in Cygnus and Mizar in the Big Dipper, and several Messier objects. I spent two hours comet hunting (CN3 is the code name I give to that project.) I also participated in the group comet search CN1, looked for Aurora Borealis, and observed three variable stars for the first time.

OBSERVING THE OBJECTS IN THE CATALOG

Now we are ready to choose our deep sky objects for observing. The following chapters arrange the objects by season, with comments on how easy they should be to observe. The Levy list numbers are based on when I first recorded each object and not on how easy to observe they are. They are not sequential in this list. However, at the end of chapter 14 there is a version of the catalog in order of position, in right ascension, in the sky.

The objects in the catalog are numbered mostly in order of when I first observed them. Why are they not presented in that order in this book?

The catalog is a personal list of objects. To turn it into a book that would be useful, I have selected about a third of the 337 objects and ordered them in chapters that take us out in increasing distance from Earth. Within each chapter the objects are described in the order they were added to the catalog. In most, but not all, cases this order would be chronological. I might have added an object after seeing it (sometimes several times!) in the course of my comet hunting, even though I might have found it deliberately during an earlier observing session. For example, I included M15 on August 23, 1966, when I chanced upon it. However, I first saw the globular in 1964. The full catalog, in its original order, can be found in chapter 14.

Each object begins with some factual information. What does that mean?

    Object number in catalog, and name:

To use the first entry as our example, L297 means that it is the 297th object I have cataloged. Please remember that these "Levy" numbers are personal designations (or in the star charts, "L" for "Levy") that I used when I decided to add a particular object to my list. I do not intend that you should start referring to NGC 1931 as Levy 1. That object is properly known as NGC 1931. Generally, my own "Levy" catalog numbers are in order of when I first saw them, but in some cases, like this one, I added them later. The object's official names, such as 47 Ursae Majoris, V Hydrae, NGC 6907, or M31, appear next.

What the object is:

This line defines the nature of what we are discussing.

The object's position in the sky:

An object's position in the sky is defined by right ascension and declination. Right ascension, denoted by the Greek letter α

(alpha), is a projection of longitude into the sky, expressed in hours, minutes, and tenths of a minute. It goes around the sky in twenty-four hours. Declination, expressed by the Greek letter δ (delta), is a projection of latitude into the sky, expressed in degrees and minutes. If objects are north of the celestial equator, they have positive (+) declinations; southern declinations are (-).

Precession:

Because the world has a slow wobble (called precession), an object's position changes in the sky with time. Astronomers list positions with respect to particular epochs, such as epoch 1950.0 and epoch 2000.0. The positions we use are the standard positions for 2000.0. These positions will probably not be changed in the literature until 2050.

The magnitude scale:

Our system of magnitudes dates back to Hipparchus, the second-century BCE Greek astronomer who divided the stars into six brightness groups. The twenty brightest stars were assigned first magnitude and the faintest stars sixth magnitude. By 1856 Norman Pogson of Radcliffe Observatory quantified these classes; a first magnitude star is one hundred times brighter than the faintest star visible without a telescope on a clear, moonless country night; a second magnitude star is two and a half times fainter than a first magnitude star; a third magnitude star, in turn, is two and a half times fainter again.

Vega, the brightest star in the Summer Triangle, is a zero magnitude star. Pogson defined Polaris, the North Pole star, as being second magnitude. Most of the stars in the Big Dipper are also about second magnitude, and most of the stars in nearby Cassiopeia are a magnitude fainter.

Best seen:

This provides an indication of what season is best for viewing a particular object. If the object is observable under a light-polluted city sky, the words "city sky" are included.

David with Miranda, the telescope with which he has discovered seven new comets. The telescope is pointed toward Leo; Jupiter is the brightest object, and the pyramid-shaped Zodiacal light can be faintly seen.
Five-minute exposure by Wendee Wallach-Levy.

Telescopes used:

Finally, each object has some personal description. In this section, I invite you to share my own experiences observing a particular object. In the description I will occasionally refer to telescopes I have used by their names. The ones most frequently used are:

Echo:	My first telescope, a 3.5-inch f/11 reflector
Pegasus:	8-inch f/7 Cave Optical Co. reflector; discovered one comet
Minerva:	6-inch f/4 Optical Craftsman reflector
Miranda:	16-inch f/5 home-assembled reflector; discovered seven comets
Cupid:	3.5-inch f/11 Questar reflector
Ophelia:	8-inch f/1.5 Celestron Schmidt Camera
Clyde:	14-inch f/2.2 Celestron Schmidt-Cassegrain used at prime focus

With this information, let us now begin our tour of the night sky.

PART 2

THE BEST AND THE BRIGHTEST

Between two worlds Life hovers like a star,
'Twixt Night and Morn, upon the
    horizon's verge.
How little do we know that which we are!
    How less what we may be! The eternal surge
Of Time and Tide rolls on and bears afar
    Our bubbles; as the old burst, new emerge,
Lashed from the foam of ages; while the graves
Of empires heave but like some passing waves.
    —Lord Byron, *Don Juan* 15.99, 1823

The objects presented here are arranged in chapters that represent increasing distance. Generally within each chapter, however, the objects appear in the order that I first put them into the catalog.

We could therefore imagine that this whole book represents a journey, from the closest worlds to the farthest galaxies. Thus, each chapter will begin with a synopsis of the journey to be taken. To measure distances, we'll use the unit called a light-year, the distance light and other forms of radiation travel in a year, about six trillion miles. Light takes eight minutes to travel from the Sun to Earth, and more than four years to travel from the nearest star, Alpha Centauri, to Earth.

CHAPTER 4

INTERESTING STARS

DISTANCES:
FROM EARTH TO HUNDREDS
OF LIGHT-YEARS AWAY

If the stars should appear one night in a thousand years, how would men believe and adore; and preserve for many generations the remembrance of the city of God which had been shown! But every night come out these envoys of beauty, and light the universe with their admonishing smile.

The stars awaken a certain reverence, because though always present, they are inaccessible; but all natural objects make a kindred impression, when the mind is open to their influence.

—Ralph Waldo Emerson, *Nature*, 1836

I shall watch from the house where some have felt peace and
hope that in your sky there are some bright stars.
—Peggy Bond Church, *The House at Otowi Bridge*, 1959

OUR JOURNEY BEGINS

As we leave Earth, we gain speed rapidly as we rush past the
Moon, and then Mars, a planet that might have once harbored
simple forms of life. We pass by the many asteroids as we move
outward to Jupiter, the king of the planets. We bid Jupiter
thanks for being the cosmic vacuum cleaner—if Jupiter's
gravity didn't change the orbits of so many comets, Earth
would still be a sitting duck for major cosmic impacts every
century. Jupiter itself is bearing some wounds, in the form of
elevated levels of carbon monoxide, from its collision with
Comet Shoemaker-Levy 9 in 1994. We move on past Saturn's
magnificent rings, Uranus tilting over on its side, and Neptune.
We then say farewell to the solar system as we pass Pluto and
Charon, two icy worlds at the edge of the solar system.

We continue moving out, beyond the point where the
Sun's outward blowing wind of energy encounters the winds
from the other stars. This is the heliopause, the true border of
the solar system. By now we have reached a velocity unimag-
inable to science—we can navigate over many light-years in a
few seconds through the magic of a telescope. Perhaps some
day we will find that Einstein-Rosen bridges, or wormholes
through space and time, can be made to last more than a
microsecond and can be used as subway tunnels to other
places. Until then, however, we'll just have to travel at the
speed of our imagination.

The closest star to the Sun is a system of three stars called
Alpha Centauri. We are four light-years from home. As we pass
it, we then move outward, rapidly increasing our distance from

Earth. We begin our list with an "object" very close to home and then move outward to the stars. (Some of the deep sky objects discussed in this book have unknown distance values, so not every object will have its distance from Earth listed.)

Levy 335   Gegenschein
The counterglow, or dust in solar system
First seen: August 20, 1966
Position (2000.0): variable
Best seen: fall; needs very dark sky and covers too much sky to be viewed
    through a telescope
Magnitude: about 5

We begin our journey with an "object" that consists of dust within our solar system. German for "counterglow," the Gegenschein is opposite the Sun in the sky, shining weakly as a large, faint oval-shaped glow about 10–15 degrees (20–30 moons) wide. It consists of dust grains in the plane of the solar system, seen by reflected sunlight. These grains of dust come from comets as they cruise by the Sun on their way through the solar system. The particles are small, perhaps a fiftieth of an inch in diameter, and are separated from each other by about five miles.

The Gegenschein is part of our view of solar system dust that extends around the sky. On very dark nights, the oval patch extends into the zodiacal band that crosses the sky to the east and west and widens near the Sun into the Zodiacal Light. It emits a teepee-shaped triangular glow that rises from the western horizon after dusk and the eastern horizon before dawn. (The Zodiacal Light is best seen on winter evenings and fall mornings in the Northern Hemisphere.)

I include the Gegenschein as something worth seeing because, on a clear, dark autumn night, when it is high in the sky and away from the Milky Way, it is the largest "object" in the sky! However, it is very difficult to see.

OF MENTORS

From my observing session log, August 20, 1966, Session 1597: "Mr. Houston indicated its approximate position. He said he wasn't sure if he saw it or not. I thought I might have seen something. He asked me to point out the center of the glow I thought I saw, with a flashlight beam. My glow coincided with his exactly. An hour later it was on the meridian and a bit brighter."

Back in the 1960s we used to go to an annual Deep Sky Wonder Night in Vermont. What really made this event special was the presence of Walter Scott Houston, the longtime author of the Deep Sky Wonders column in *Sky & Telescope* magazine. To meet the man in person, with his wife and family, was an extraordinary privilege. I loved his wry sense of humor and the continual stories he told us about his observing experiences. I'll never forget his tale of spotting Comet Mrkos. During the summer of 1957, this new bright comet appeared in the twilight evening sky. Scotty was enjoying an after-dinner puff on his pipe when Clifford Simpson, who was with him at the time, asked him why he had not pointed out the comet. Scotty assumed that Simpson was speaking of an earlier comet, Arend-Roland, which had been bright some weeks before but was now faded. Not missing a puff, Scotty said that there was no comet. "But there is a comet there!" Scotty looked up and saw one of the brightest comets of his life. He tossed the lit pipe into his jacket pocket, darted to his telescope, yanked it with one hand, focused it with the other, and peered at Comet Mrkos as Simpson saw the eruption of smoke from Scotty's back pocket. "Scotty, you're on fire!" he hollered. Ripping off his burning jacket and throwing it on the grass, Houston was about to return to his telescope when the grass also caught fire. Finally they smothered the flames, just in time to lose the comet behind a tree.

Scotty was a master storyteller, but he told me this one so many times that I believed it really happened. He was a mentor with a lot of wisdom to offer. One night as amateur astronomer Rik Hill and I sat talking with him about a new book, Rik laughed that the book might already be out of date. "No book which faithfully represents the passion of its author can ever really be out of date," Scotty said. For me, Scotty's grandest lesson came during Deep Sky Wonder Night that August night in 1966. Only eight months into my comet search, I enthusiastically explained to him my plans to scan the sky as we stood inside the farmhouse kitchen. He pondered my words for a moment, took another puff on his fabled pipe, and asked, "What's the sky like now?"

"Dark and clear," I smiled.

"Then how do you expect to find a comet while we're standing inside chatting?"

Scotty also introduced me to the Gegenschein that night; if you can see it, Scotty pointed out, it is the mark of a beautiful, dark night. Years later, my wife, Wendee, glimpsed what she thought was a dull glow. "David," she asked, "is that the Gegenschein?" I then gave her Scotty's test, asking her to compare her view of it with mine. Wendee's enjoyment in her first observation reminded me of mine back in 1966, and also of the wise mentor Walter Scott Houston.

Levy 69P   Wendee's Ring, LW   J2204.3+4508
Asterism, or chance grouping of stars
(This style of designation is often used in astronomy now. Note that the
    numbers correspond to the position in the sky.)
First seen: January 2, 2000
Position (2000.0): α 22 04.0   δ +45 09
Magnitude: about 14
Best seen: in autumn; needs dark sky

Levy 69P, Wendee's Ring asterism.

One of the strange things about living in the world is that it is only now and then one is quite sure one is going to live for ever and ever and ever. One knows it sometimes when one gets up at the tender solemn dawn-time and goes out and stands alone and throws one's head far back and looks up and up and watches the pale sky slowly changing and flushing and marvelous unknown things happening until the East almost makes one cry out and one's heart stands still at the strange unchanging majesty of the rising of the Sun—which has been happening every morning for thousands and thousands and thousands of years. One knows it then for a moment or so. And one knows it sometimes when one stands by oneself in a wood at sunset and the mysterious deep gold stillness slanting

through and under the branches seems to be saying slowly again and again something one cannot quite hear, however much one tries. Then sometimes the immense quiet of the dark blue at night with millions of stars waiting and watching makes one sure; and sometimes a sound of far-off music makes it true; and sometimes a look in someone's eyes.
—Frances Hodgson Burnett, *The Secret Garden*, 1911

Comet hunting is my secret garden. When I come across something unique in the sky, I get the same feeling that Burnett describes so well in her story. I got that feeling when "Wendee's Ring" of faint stars was discovered during the course of a photographic survey of the sky that Wendee and I made.

January 2, 2000, was a day with cirrus clouds, but with the setting Sun, the clouds seemed to disappear with a promise of a beautiful night. We walked out to our observatory, a 20-by-32-foot building with a retractable roof that stands in the midst of our backyard. We pushed the roof along its rollers until it revealed the darkening sky. Wendee and I took a series of wide-angle photographs, each one covering a 10-degree square area of sky. After the films were processed, we scanned them the next day using a device called a stereomicroscope. Each pair of films showed an identical patch of sky, but if a comet or an asteroid should be moving through that field, it would appear to float atop the starry background. It is an elegant way to search for comets or anything else of interest that the sky might have to offer.

There were no new comets in those pictures. But while I was scanning one of the pairs of films, I encountered an incredibly beautiful ring of 12 and 13 magnitude stars that was barely visible above the background of sky. The 7 arcminute-wide ring (i.e., 7 minutes of declination) was open at its southern side. The following day I contacted Brent Archinal, an astronomer at the United States Naval Observatory who is an expert on the clusters, associations, and chance groupings or asterisms of stars in the Milky Way. Though I thought this must be one of the several known chains of stars that are scattered across the sky, Brent surprised me with the news that the ring had not been documented before.

Officially known as Levy-Wallach J2204.3+4508, according to its position in the constellation of Lacerta, the nickname of this interesting chain of about forty mostly 14 magnitude stars is Wendee's Ring.[1] The ring was an ethereal sight on those discovery films. This is no comet moving inexorably toward Jupiter, no asteroid speeding past the Earth. It is a group of distant suns frozen in the sky that I would never have seen had I not searched for comets, and it is now part of the secret garden that is my personal sky.

Levy 70P   Equuleus S, LWJ2108.8+0620
Asterism
First seen: December 25, 2000
Position (2000.0): α 21 09.0   δ +06 18
Magnitude: about 12
Best seen: in autumn; needs dark sky

A few hours before Wendee and I set up our observatory to take comet search photographs on Christmas night 2000, our observatory was a busy place. We had invited all our neighbors to see a partial eclipse of the Sun. We thought that the eclipse would be the highlight of our day, but when I scanned our films the next day, I found this S-shaped asterism. Like Wendee's Ring, it is likely that its stars are at different distances from us and that they are not physically related.

The brightest star in the S is HD 201331 (the Henry Draper Catalog). It includes about twenty-five stars of similar brightness.[2]

Levy 70P, Equuleus S.

Levy 71   Nanette's River, LW J2340.6+5618
Asterism
First seen: May 3, 2001
Position (2000.0): α 23 40.6   δ +56 18
Magnitude: about 9
Best seen: in autumn and winter; rich and beautiful even in city sky

At about midnight on the mild spring night of May 3, 2001, I walked into the kitchen for a break in my observing session. Our daughter Nanette was awake in the kitchen also, so we chatted a while. I told her how everything seemed to be going so well; Miranda, my 16-inch telescope, was working beauti-fully, and the sky was pristine. It turned out that Nanette

Levy 71, Nanette's River.

wasn't sleeping any better than I was that night, and each time I came inside, there she was!

I didn't come across any comets that night, but around 3 AM I did spot a unique and beautiful chain of stars—an asterism—that wound its way across more than a degree of sky. When I came in around 3 AM for another break, Nanette was up again, and I excitedly told her about it. Just before I went out, I looked at her and said, "If no one else has described this asterism, I'd like to call it Nanette's River."

I submitted the description and proposed name to Brent Archinal, who was at that moment in the midst of a massive project updating the identities and positions of the many open clusters and interesting asterisms that observers have described. It turned out that, indeed, this grouping had never been reported before.

When Nanette came for another visit in the fall of 2002, this time with Summer, her daughter, and her new son, Matthew, she got to look through a telescope at her new "river." It was a joy to see how ecstatic she was about her own piece of celestial real estate.[3]

Levy 72   V Hydrae
Red variable star
First seen: November 22, 1984
Position (2000.0): α 10 51.5 δ –21 10
Magnitude: varies 6.6–9.0
Distance: 20,000 light-years
Best seen: in spring; observable in city sky
A Sun-like star that has at least two planets orbiting it in almost circular orbits.

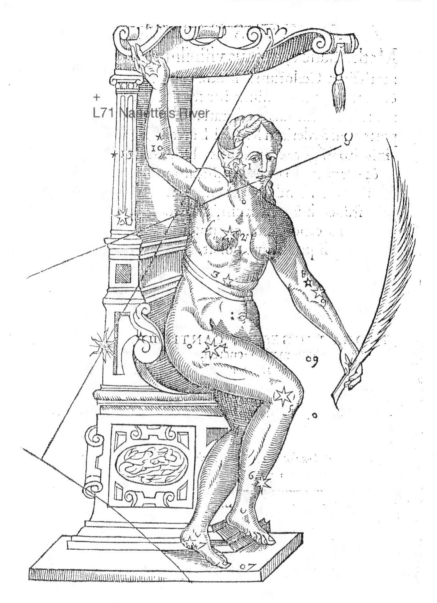

Nanette's River's position on a drawing of Cassiopeia originally published in 1573 in Thomas Digges's Alae seu scalae Mathematicae. *The map was intended to show the position of the supernova of 1572; I have added a "+" to show the approximate position of L71 Nanette's River. I thank the Cambridge University Library for its help in obtaining this image.*

During my first hour of searching after finding my first comet in 1984, this extremely red star made its appearance in my telescope. It is one of the reddest stars I've ever seen, rivaling Mu Cephei and R Leporis. Especially since I wasn't expecting it, this star imprinted itself on my eyes and in my memory. It provided a lesson that one should not just look for the prettiest galaxy or cluster; a single star, if its color is unusual enough, qualifies as something worth spending some time to find and admire.

What is a variable star? Stars share something with people —they seem to have moods. Sometimes the light output of these stars changes. When a star is young, it might flicker with the intensity of a rebellious child, unable to decide on its future course. As it gets older, the star tends to acquire the dignity of middle age, staying out of trouble and perhaps bearing the responsibility of lighting and heating a family of planets.

Old age does strange things to a star. Now a red giant, it may begin varying again, but this time to the slow and measured beat of an aging heart. V Hydrae is such a star, and so are R Leporis and SS Virginis. Or the star may become unpredictable, varying over time with graceful but completely irregular pulses.

Levy 100   Beta Persei, Algol
Eclipsing binary variable star
Added to catalog: March 23, 1997
Position (2000.0): α 03 08.1 δ +40 58
Magnitude: varies 2.3–3.5
Distance: 92.8 light-years
Best seen: in fall and winter; easily seen in city sky

Algol has attracted my curiosity ever since I read somewhere that a British amateur astronomer was so mesmerized by the miracle of this "winking star" that whenever it faded, he set up his refractor telescope on London Bridge to show it to as many people as possible.

Actually a triple star system, Algol's name means the Demon, or the Blinking Demon, from the Arabic "Ra's al Ghul" for the "Demon's Head," according to astronomical historian Richard Hinckley Allen. Allen believes the name does not necessarily come from an ancient discovery of the star's variability, but that it derives from the Greek astronomer Ptolemy. "Al Ghul literally signifies a Mischief-maker," Allen continues, "and the name still appears in the Ghoul of the *Arabian Nights* and of our day." The ancient Hebrews in Babylon thought of Algol as Lilith, Adam's supposed first wife. In this fascinating Talmudic story, Adam separated from this first wife and their demon children before Eve was created for him. Alternatively, the name "Rosh ha Satan," or Satan's head, was used as well.[4]

THE JOHN GOODRICKE AWARD

Although Geminiano Montanari of Bologna described its variability around 1667, the first precise determination of Algol's period of light variation was fifteen years later by John Goodricke, a teenage amateur astronomer whose life was inspiring. (In fact, over the years I have encouraged astronomy clubs to present a John Goodricke Award to any of their youngest observers, to help foster the growth of their interest in astronomy and to turn a childhood fancy into a lifelong joy. The reason I named this award after Goodricke is that he led a special and inspiring life.)

John Goodricke was a deaf mute who lived in a time when these conditions were often equated with stupidity. However, Goodricke's enlightened parents did not believe this and sent him to a school in Edinburgh, whose headmaster had developed a process to teach deaf children to speak and to think logically. Meantime, his parents moved to York, England, by fortunate accident just a few houses from the Pigott family, in which Nathaniel and his son Edward were astronomers. When John Goodricke returned to live with his parents, he had developed at least a theoretical interest in astronomy, and he quickly became friends with Edward. With Herschel's discovery of Uranus taking place at the same time, the two young men must have had much to consider.

On November 15, 1781, Edward independently discovered a new comet "with a small nucleus and coma near the neck of Cygnus." This was actually Comet Mechain, 1781 II, which had been discovered five weeks earlier by the French observer whose name it bears. By the time Pigott observed this comet, it was a conspicuous object of at least 4 magnitude, with a tail four degrees long. Seeing this comet inspired Goodricke to begin a diary of personal astronomical observations, and he described his friend's discovery on its opening page.

When Edward suggested a few months later that he begin a search for new variable stars, John loved the idea. On November 12, almost a year after the Cygnus comet, John observed a sudden drop in Algol's brightness. Edward was certain that Algol was variable, but neither man had any idea that the change would come so quickly. Six weeks later, on December 28, the two friends observed the variation again. In 1782 they determined Algol's period of variation as sixty-nine hours. As for why it varies, John Goodricke offered two possibilities. One was that the star was partly covered by large dark markings or spots that would, through rotation, cause the drop in brightness. The other theory was that a companion object revolved around the star. Not until more than a century later, during the 1880s, did theoretical work by E. C. Pickering and observations with a spectroscope by H. C. Vogel finally give voice and hearing to the work of the eighteen-year-old Goodricke.

The two young men continued their work. On the night of September 10, 1784, they independently discovered two of our best known variable stars. As Edward was detecting variability in Eta Aquilae, John observed a brightness change in Beta Lyrae. A month later, Delta Cephei revealed its secret to Goodricke's astute eye. Thus, by the age of nineteen, Goodricke had discovered the variations of three stars. In early April 1786, after a series of Delta Cephei observations on what must have been cold nights, he became ill, possibly with pneumonia. He died on April 20. Edward was shattered, losing his joy in observing. He had been away from York at the time and simply continued his wanderings for over a decade. Fortunately he did recover his energy and went on to discover the variations of R Scuti and R Coronae Borealis.[5]

ECLIPSING BINARY STARS

One of Goodricke's theories was correct: Algol indeed had a companion star. Almost half the stars in the sky are binary systems, containing at least two stars revolving around each other. With so many dancing partners around us, some small percentage should pass each other exactly in our line of sight, so that they appear to be getting in each other's way. One star could pass either partially or fully in front of its companion. In either case, when the fainter star passes in front of the brighter, we could observe a drop in total brightness. We know of over forty-seven hundred eclipsing binaries, composed of stars too close to each other to be seen separately through a telescope. Their behavior, not their appearance, gives away their secret.

Algol varies because its two main stars, Algol A and Algol B, revolve around each other on our line of sight, so that B eclipses A from time to time. In addition to the close binary AB, however, there is a third companion called Algol C. Algol A is a bluish white main sequence star with a mass 3.6 times that of the Sun. B is a giant star, an orange-red sun that has been severely changed by its tidal interaction with A. Tidal forces from A have distorted B into the shape of a teardrop. B is almost as massive as the Sun but has more than three times its diameter. The last component, Algol C, is a bluish white dwarf star, 1.5 times the Sun's mass but four times brighter. It orbits the system every 1.9 years.

Can we say that eclipsing binaries are not really variable stars, that they vary only because of their geometrical alignment? There probably is some minor real variation going on in some eclipsing binaries. Algol A and B are so close that the gravitational pull and the magnetic field interaction of one on the other is probably sufficient to induce real, though minor, change in the physical brightness of each star. However, the variation we can see with our eyes is due entirely to the eclipses.

The beauty of this famous star is that its whole eclipse is visible to the naked eye. For most of its cycle—almost fifty-nine hours—the star stays at magnitude 2.3. (After 29.5 hours, A passes partially in front of B, causing a barely noticeable drop of a twentieth of a magnitude). After the full fifty-nine hours, the ten-hour eclipse begins anew. In the space of about five hours, B passes in front, and the star fades to magnitude 3.7. The minimum, during which B is directly in front of A, lasts for about twenty minutes, and the star brightens up again in another five hours.

Even though the eclipses take place frequently, you may have to wait before you can spot one at a convenient time. Some take place during daylight hours or begin late at night. Check the predictions that are published each year in the *Observer's Handbook* of the Royal Astronomical Society of Canada, or each month in *Sky & Telescope* magazine, for the time of the next eclipse that is convenient for you. Remember that the predictions in the *Observer's Handbook* are in universal time, which is five hours ahead of Eastern Standard Time, or four hours ahead of EDT.

Levy 148   R Leporis
Red variable star
Cataloged: October 31, 2002
Position (2000.0): α 04 59.6   δ −14 48
Magnitude: about 7.3–9.8
Distance: about 1,500 light-years
Best seen: in winter; observable in city sky

In my early years of stargazing, I longed to see the little rabbit that hung in the sky just south of Orion—Lepus the Rabbit. It was so far south that it couldn't quite hop above the treetops over the southern horizon of our Montreal home. One of the reasons I longed to see Lepus was the strange star it contained—Hind's Crimson star, also known as R Leporis. It was supposed to be as red as a drop of blood. When I finally saw it at a faint 11 magnitude, it looked only mildly red.

As R Leporis brightened over that winter, however, its red color deepened noticeably. When I began my second season of observation, it greeted me as a true blood red, 8 magnitude star. Named by J. Russell Hind, a nineteenth-century British astronomer, this star changes in magnitude from 7.3 to 9.8 and back again over 420 days, so over a period of a year you can see it fade or brighten.

Levy 157   The Cane, LW J1948.2+3743
Asterism
First seen: November 4, 2002
Position (2000.0): α 19 48.2   δ +37 43
Magnitude: about 9
Best seen: in summer; rich and beautiful even in city sky

I found this beautiful cane-shaped stream of stars while comet hunting on November 4, 2002. It is a stream of about twelve stars, about 33.5 arcminutes long, roughly centered on the 7.2

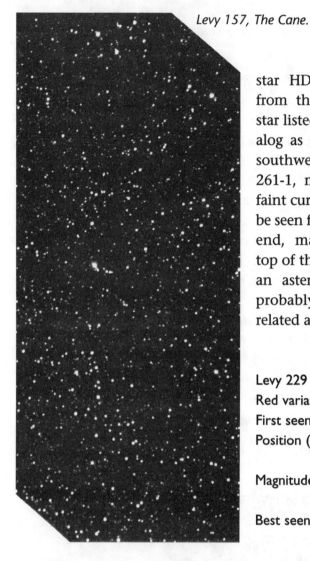

Levy 157, The Cane.

star HD 187374. It goes from the 10.6 magnitude star listed in the Tycho catalog as 3136-343-1 to the southwest to TYC 3136-261-1, magnitude 11.8. A faint curve of stars can also be seen from the southwest end, making the curved top of the cane. Again, it is an asterism; its stars are probably not physically related as a cluster.[6]

Levy 229   SS Virginis
Red variable star
First seen: January 1, 2003
Position (2000.0): α 12 25.2
                        δ +00 46
Magnitude: 6.0–9.6,
                varying over a year
Best seen: in spring; observable in city sky

Not far from this unusually red variable star is the brightest quasar in the sky, 3C 273 (L230). The star is in our own galaxy, but since the quasar is at the limits of the Universe, our discussion of it will have to wait until chapter 11.

Levy 241   AA Ceti
Double star
First seen: January 21, 2003
Position (2000.0): α 01 59.0   δ −22 55
Magnitude of brighter star: 6.2
Best seen: in autumn; observable in city sky

Between a third and a half of all the stars in the sky are double stars, that is, two stars revolving around each other. There also are triple, quadruple, and other multiple star systems.

Discovered by William Herschel in 1822, AA Ceti is actually a triple star. The two brightest members of the AA Ceti family are beautiful through a small telescope. They are separated by 8.4 arcseconds of declination at an angle of 304 degrees; that means the fainter star is northwest of the brighter one. (If it were due west, it would be 270 degrees, or due north it would be 0 degrees.) The brighter star, AA, has a small third star orbiting nearby. We can't see it, but the star gives its presence away because it passes directly in front of the brighter star, causing its brightness to drop by about half a magnitude.

Levy 297   47 Ursae Majoris
Single star
First seen: May 1, 1964
Position (2000.0): α 10 59.7   δ +40 24
Magnitude: 5.0
Distance: about 42 light-years
Best seen: in spring and summer; observable in city sky
A Sun-like star that has at least two planets orbiting it in almost circular
     orbits

This next stop on our journey is a single star that doesn't look like much through a telescope. When I think of this remarkable star, though, I am reminded of a summer evening over dinner in 1960, when Dad shared with us a story he had read when he was a youngster. It was *Cole of Spyglass Mountain*, Arthur Preston Hankins's novel about a boy whose love of the sky led him to observe Mars through his small homemade tele-scope.[7] The novel included some pretty violent moments, including gunshots being fired in the observatory, but it ended with Cole finding evidence of life on Mars one night and becoming an instant celebrity. "Now that you're interested in astronomy," Dad said, "if you ever find that book in your wan-derings through astronomy libraries, I'd love to read it again."

Five years later I began my long search for comets. As the years went by and Dad grew older, he would often ask if I had found *Cole of Spyglass Mountain*. Then with the onset of Alzheimer's disease, his memory began to fail. It wasn't long before he began to forget names, people, and once even forgot that I was his son. But somehow he never forgot that mar-velous tale. When I finally found my first comet in 1984, I felt as though I had rewritten *Cole of Spyglass Mountain* just for Dad. But I think that he was too ill to appreciate it, and he died only a few months later. When my mother and I tried to tell him about it, he did smile. I hope he understood.

Life on other worlds was a subject about which Dad was always excited. I think he would have been interested to learn of the story of 47 Ursae Majoris, some 42 light-years away. That sun, just below the bowl of the Big Dipper, is faintly visible to the naked eye. Four years after Dad told me about *Cole of Spy-glass Mountain*, I was assigned a small area of the sky to search each night with binoculars for possible comets. I have been familiar with 47 Ursae Majoris since I first looked at it on May 1, 1964.

As a freshly minted member of the Royal Astronomical Society of Canada, I became part of a group effort to search the sky for comets and novae. Each of us was assigned a small region of sky to watch; mine was area 377, just south of the Big

Dipper's Bowl. I got to know that area very well over the years—the pattern of its stars and their different magnitudes. One of its brighter stars is 47 Ursae Majoris. So I was delighted beyond measure when in 1996 Geoffrey Marcy and Paul Butler, at the University of California at Berkeley, announced that a planet twice the size of Jupiter is orbiting that star. They called it 47 Ursae Majoris b.

Five years later, Marcy, Butler, and Debra Fischer, also from Berkeley, had even better news. There wasn't just one planet: 47 UMa is a solar system with at least two worlds. The new one is called 47 Ursae Majoris c. More like our own than any other seen to date, the planets orbit that sun in almost circular paths.

Why is this important? From 1930, when Pluto was discovered, to 1992, we knew of no new major planets—anywhere. Since then our known roster of planets, and whole solar systems, has grown exponentially. As of the beginning of 2005, ninety-eight stars had known systems of planets, with a total of 134 planets. That number rises every month. However, most of these planets circle their suns in wildly elliptical paths, rushing out to great distances and then closing in on them. If our solar system had planets like these, our hapless Earth would have long ago been destroyed in a collision with one of them. In this new solar system of 47 Ursae Majoris, the worlds orbit peacefully in almost circular orbits much like Jupiter and Saturn, leaving plenty of room for safely orbiting smaller worlds that could have life.

Forty-seven Ursae Majoris is the type of sedate system astronomers have been looking for. Wouldn't it be wonderful, even if unlikely, if someone on one of those worlds—or on a moon orbiting one of those worlds—would be looking back at us, asking the same questions we ask, wondering what we wonder: Are we alone?

Perhaps that distant person would be lucky enough to have a father with the same taste for literature that my father had. A few years after his death, I finally located a copy of *Cole of Spy-*

glass Mountain. As a young boy, Joshua Cole was taken to a "house of refuge" where he lived with other boys, each given a number. Coles's was "number fifty-six thirty-five." After I finished the story, I contacted Bobby Bus, the discoverer of the still-unnamed asteroid 5635, and suggested that it be named in honor of my father's fictitious young friend. The proposal to the International Astronomical Union's appropriate committee was enthusiastically approved, and now a real-world 5635 Cole honors Cole of Spyglass Mountain.

Levy 329   T Coronae Borealis
First seen: 1966
Position (2000.0): α 15 39.5 δ +25 55
Magnitude: 10.0, exploding to 2
Distance: about 2,000 light-years
Best seen: in spring and summer; observable in city sky

This strange star is a nova, a star undergoing an explosion. Its first recorded outburst occurred in 1866, when suddenly it rose to 2 magnitude and then slowly faded. With the increasing popularity of variable star observing in the early twentieth century, observers naturally turned to the old novae just to see what they were doing, particularly this puzzling star that seemed to vary irregularly around its 10 magnitude minimum. Then, in February 1946, this nightly observing paid off when T Coronae erupted again, bursting overnight to 2 magnitude.

Over many years, I have watched as T Coronae sleeps fitfully, ranging irregularly in brightness from 9.9 to 10.2. Astronomers have even recorded flickering by 0.1 magnitude over several minutes. T is a binary star, one of whose members might be a large red semiregular variable that causes the slow oscillations. The other component is a smaller and bluer star. That star does the rapid flickering and nova outbursts. The two stars orbit each other in slightly less than eight months.

T Coronae Borealis is the best known example of what we call a recurrent nova. Although we suspect that all novae recur eventually, their outbursts are separated by periods as long as ten thousand years. When a star like T Coronae Borealis erupts, it undergoes a nuclear chain reaction on the surface of the small star. A combination of high temperature and high pressure leads to an explosion much like that of a hydrogen bomb, but on even a larger scale. We know of only five recurrent novae, including T Coronae Borealis. The other four are RS Ophiuchi, T Pyxidis, V1017 Sagittarii, and U Scorpii. The recurrences are not regular and, in the case of RS Ophiuchi, can be as frequent as nine years. Observing these stars is a worthwhile pursuit for amateurs.

OF STARS AND FRIENDS

If my own early observations of T Coronae Borealis made me wonder about a 10 magnitude star that could, within a few hours, brighten to 2 magnitude, the great comet hunter Leslie Peltier brought the star's personality home. As he became seriously interested in exploding stars like T Coronae, he decided to watch them every clear night, in the hope that he would catch them exploding a second time. T Coronae Borealis, with its small fluctuations, really intrigued him. "For more than twenty-five years," he wrote, "I looked on it from night to night as it tossed and turned in fitful slumber. Then, one night in February 1946 it stirred, slowly opened its eyes, then quickly threw aside the draperies of its couch and rose!

"Full eighty years had passed since last the star had shattered the symmetry of the northern crown. And where was I, its self-appointed guardian on that once-in-a-lifetime night when it awoke? I was asleep!"

Leslie went on to write how the alarm clock had awakened him that morning but that he decided to return to the warmth of his bed and bypass the frigid night outside. "And thus I missed the night of nights in the life of T Coronae. . . . I still am watching it, but now it is with a wary eye. There is no warmth between us any more."[8]

In these words Peltier gave the star and himself the relationship that defines friend or family. In reality, T Coronae exploded some two thousand years before 1946, and the light of that outburst traveled the distance to reach the covered lens of Peltier's telescope on the one night he decided to take care of his health and not brave the early morning cold. But in another view of reality, the sense of family between star and observer is quite real, and I understand and share the great observer's feelings about the stars he spent his time watching. T Coronae has not disappointed me, at least not yet. In this personal sense then, I add T Coronae Borealis to my own list, in honor of my wise friend Leslie Peltier.

NGC 663 (Levy 379), an open cluster photographed by the author using Clyde, a 14-inch Schmidt-Cassegrain telescope with CCD and Starizona Hyperstar.

CHAPTER 5

OPEN STAR CLUSTERS
DISTANCES: HUNDREDS TO THOUSANDS OF LIGHT-YEARS AWAY

Pistol: We have seen the seven stars.
>   —William Shakespeare,
>   *2 Henry IV*, 2.4.177, 1597

Fool: Thy asses are gone about 'em. The reason why the seven stars are no more than seven is a pretty reason.
Lear: Because they are not eight?
Fool: Yes, indeed: thou wouldst make a good fool.
>   —Shakespeare, *King Lear*, 1.5.33-36, 1605

Having stopped by some interesting stars, we now travel deeper into the galaxy, past the

Pleiades, a star cluster containing several hundred stars that, bound gravitationally to each other, move through space together and evoke enough romance to have appeared in English drama poetry for hundreds of years. The Pleiades is a gorgeous grouping of hot blue stars, some surrounded by gas. It is so beautiful that even though its stars are not the brightest in the sky, almost everyone who looks up catches a glimpse of them. Even first-time observers ask about "that fuzzy grouping over there in the east."

Probably all stars, including our Sun, were born in clusters. The cluster of our Sun must have been a spectacular sight indeed five billion years ago. Its stars have long since left their cosmic nest and have spread throughout the galaxy. We have no way of knowing which of the distant stars we see in the night sky are sisters of the Sun. Could 47 Ursae Majoris be one of them?

We are passing through an open cluster right now. It is a loose association of stars called the "Ursa Major moving cluster"; it is also known as Collinder 285. The nature of this cluster was first noted in 1872 by astronomers Richard Proctor and William Huggins. (We call it a moving cluster because the rates and directions of motion of its stars have been measured, and all the stars are moving in the same direction and at the same velocity.) Most of the Big Dipper stars are part of it. The center of the cluster is only seventy-five light-years from us.

How do we know that so many far-flung stars belong to a single cluster? We can measure what we call their proper motions, which give us an idea of the speed and direction in which the stars are moving. So when astronomers found that a large number of stars in our vicinity of space were moving with a common speed and direction, they concluded that these were part of a cluster. The Ursa Major moving cluster consists of some of the Dipper's bright stars, like Epsilon, Beta, Zeta (i.e., the double star Mizar), Gamma, and Delta. Alphecca and Alpha Coronae Borealis may also be members. Our Sun is not a member.

The Ursa Major cluster is surrounded by a stream of stars that shares its speed and direction, stars that were once members of the cluster but which are now gradually moving away. This Ursa Major stream includes stars from as far around the sky as Alpha Ophiuchi, Delta Leonis, Beta Aurigae, and Sirius. The stream stretches out to about one hundred light-years from the center. Obviously the stream surrounds us; our Sun is currently passing through it. We know our Sun is not a member because it is so much older than the other stars and because the member stars are passing us at a velocity of almost thirty miles per second.

What is the difference between chance groupings of stars that we see scattered all over the skies and real open clusters? Membership in a cluster depends on the stars sharing two factors with already established members: proper motion and radio velocity. In 1718 Edmond Halley discovered that Arcturus and Sirius had moved slightly in the twelve hundred years since Ptolemy had last recorded their positions; he discovered that fixed stars actually move relative to one another. These motions are known as proper motions. Most stars are so far away that fifty to one hundred years are required to show this motion. Radial velocity is the line-of-sight measure of how fast a star is moving, either away from us or toward us. If a group of stars shares distance, proper motion, and radial velocity, that would indicate that the stars are members of a true cluster. If not, then they are called an asterism.

Levy 53   NGC 752
Open star cluster in Andromeda
First seen: July 4, 2002
Position (2000.0): α 01 57.8   δ +37 41
Magnitude: 5.7
Distance: about 1,300 light-years
Best seen: in fall and winter; observable in city sky

A rich open cluster
Shapley class d; Trumpler class III 1 m

This is a very satisfying open cluster in Andromeda. Giovanni Batista Hodierna might have been its discoverer during the middle years of the seventeenth century. But if he or William Herschel, who observed and cataloged it as No. 32 in his seventh list in 1786, could have seen this cluster through a modern wide-field telescope, either man would have been really impressed. Two centuries ago telescopes had narrow fields of view, so observers looking through them could not see the majesty of the full cluster surrounded by space in a single field of view. In essence, they could not see the forest for the trees. I encountered this cluster while comet hunting with Pegasus, my Cave 8-inch f/7 reflector with Nagler 31 mm eyepiece.

SHAPLEY'S CLASSIFICATION SCHEME FOR OPEN STAR CLUSTERS

On the evening of February 9, 1963 (exactly twenty-three years before Halley's comet rounded the Sun), I learned all about how the great Harlow Shapley classified open star clusters on the basis of how concentrated they appeared through a telescope. During our evening study period at the Jewish National Home for Asthmatic Children, the telephone rang and our houseparent came to get me. At the other end of the line was one of the more senior members of the Denver Astronomical Society. He had heard that I was writing a book about astronomers, and he wanted to tell me about Shapley, one of the twentieth century's truly great astronomers. (There will be a discussion of Shapley's observing and his work on the distances of the globular clusters in the commentary on L195 NGC 2419, Shapley's Intergalactic Wanderer, in chapter 9.)

Over the phone that evening, I learned how Shapley and his student, Helen Sawyer, developed a scheme for open clusters and another for globulars.

For open clusters, the scheme is:

c: very loose and irregular
d: loose and poor
e: intermediate rich
f: fairly rich
g: considerably rich and concentrated

That system has since been superceded by the more comprehensive three-part method of Robert Trumpler, in which concentration is listed as:

I.   detached from surrounding star field; strong concentration toward center
II.   detached; weak concentration toward center
III.   detached; no concentration toward center
IV.   not detached from surrounding star field

The range of brightness among a cluster's stars is listed as:

1.   small range in brightness
2.   moderate range in brightness
3.   large range in brightness.

The richness of a cluster is recorded thus:

p.   poor (fewer than fifty stars)
m.   moderately rich (fifty to one hundred stars)
r.   rich (more than one hundred stars)

If the cluster has nebulosity associated with it, an "n" follows.

Levy 85   IC1396

Triple star, star cluster, and Elephant Trunk Nebula

First seen: 1966

Position (2000.0): α 21 39.1   δ +57 30

Magnitude: 5.0

Distance: about 3,000 light-years

Best seen: year round from most of Northern Hemisphere, though best
    in winter; observable in city sky

A bright grouping of stars centered around a double and a triple star

Shapley class c; Trumpler class II 3 m n

Back in 1966, while at Westmount High School in Montreal, my friend Carl Jorgensen and I were both active in observing and in encouraging others to observe; Carl's specialty was double stars. I was interested in things that *happened*, like stars that change in brightness or comets that pushed their way across the sky. But in their rich fields and colors, Carl felt the majesty of double and multiple stars, and one evening he

Levy 85, IC1396.

imparted that enthusiasm to me by showing me the field of IC 1396, a wide-open cluster that contained a triple star and a double star, Sigma 2816 and Sigma 2819. This is a stunning field of view that is available virtually year round from most of the Northern Hemisphere. Besides the triple and double, the cluster consists of several dozen stars. It is a wonderful target for star parties.

After forty years, Carl and I are still good friends. He runs the computer center for McGill University's Department of Electrical Engineering, and whenever we visit, it is as if our last meeting was yesterday.

Listed in other catalogs as Trumpler 37 and Markarian 47,[1] this cluster is believed to be about three thousand light-years away. This cluster's stars are quite young and newly formed. In fact, the Elephant Trunk Nebula works its way through the cluster; it is a cloud of gas and dust that contains the raw material for new stars.

The reason I like this cluster is that it has a star attraction: the triple Σ2816 (or Struve 2816, for the astronomer Friedrich Georg Wilhelm Struve who first cataloged it in the 1830s)— absolutely my favorite triple. Double Star Σ2819 is also quite beautiful. The rest of the cluster is very wide—about 5 degrees in diameter, with some branches of stars that trail off. Because the cluster is so large, and because it is located in a relatively rich region of the winter Milky Way, it is difficult to tell where the cluster ends.

Levy 98   NGC 6709
First seen: November 13, 1984
Position (2000.0): α 18 51.5   δ +10 21
Magnitude: 6.7
Distance: about 9,100 light-years
Best seen: in summer and fall; observable in city sky
Shapley class d; Trumpler class III 2 m

Levy 98, NGC 6709.

NGC 6709 is a very pretty open cluster consisting of some forty faint stars surrounded by a rich Milky Way field in Aquila. Always a lovely sight, on the night of November 13, 1984, the cluster helped provide the most beautiful sight I had seen in all my years of stargazing. The evening's observing followed a truncated dinner with Lonny Baker, a friend who was in charge of a lecture series, Eyes on the Universe, at the Flandrau Planetarium in Tucson, Arizona. I looked forward to the series; nothing, in fact, could keep me from this lecture.

Nothing, that is, except a clear and moonless night. Although I had by then devoted 917 hours with my eye at the eyepiece and every dime I could scrape together to further my quest, I had not had the experience of finding a comet on my own. I had moved from cloudy Montreal to the Arizona desert southeast of Tucson, and I had observed every known comet I could find.

The late afternoon clouds on November 13, 1984, promised a leisurely evening and a pleasant lecture, but as dinner went on I sensed that the clouds were getting thinner. And Lonny could see that instead of concentrating on her words, I was looking past her out the window.

"David," Lonny said, "it must be clearing up outside."

"Uh-huh," I replied absentmindedly.

"You're going to stand me up, aren't you?" she demanded. "You're going to go home and hunt for comets, aren't you?"

"Oh no!" I protested, gamely snapping to attention. "We are going to finish dinner. *Then* I am going to stand you up, go home, and hunt for comets."

Lonny laughed and said, "You'd better find me a comet tonight!"

The half-hour drive for home was shorter than thirty minutes that night as I sped along. It also wasn't the first time I did that to catch an hour or two of clear, dark sky. On one of those drives I would get a speeding ticket, in fact, though not this time.

With the telescope's slow, deliberate motion across a portion of sky, comet hunting is not like a star party, in which people line up to look at an object. On such occasions, the sky is asked to be a servant, showing off Saturn, the Moon, or some galaxy on cue. It's the opposite with comet hunting. When I start a session, I have only a vague idea of what I might find in the next hour or so as I move the telescope forward for a few minutes across a region of sky, then backward through the next sector. Whether I find a star cluster or a galaxy, a red star or a bright double star, is really up to the sky, not me. The sky is the master, my telescope the receiver, and I am the watchman.

Out in the observatory on that night of Tuesday the thirteenth, I enjoyed a most pleasant hour of comet hunting. After some thirty minutes had gone by, a faint fuzzy object appeared in the field of my moving telescope. It had the appearance of a globular cluster, I thought, and a quick check of an atlas con-

firmed my suspicion that it was NGC 6229, the third and forgotten globular cluster in Hercules. Like a fish thrown back into
the water, the cluster was gone, and I was on my way. The next
object was a planetary nebula, the remnant of an outburst in
an ancient star—interesting but not my quarry.

Next came NGC 6709. The sky was so nice, and the cluster
so striking, surrounded as it was by the rich background of the
Aquila Milky Way. Just to the south were three stars roughly in
the shape of a boomerang, and next to them was a fuzzy object
almost as bright as the cluster. It was a striking sight, the beautiful cluster and the faint fuzzy spot; my first reaction was
"Why have I never seen such a thing? It should be pictured in
all the astronomy books." Another atlas check confirmed my
growing suspicion: the cluster belonged there; the faint fuzzy
spot did not. Within a few minutes I was sure that the object
was moving very slowly in the direction of the cluster. A comet!
My heart rate soared.

It was a comet all right, but was it already known? I called
Brian Skiff, an observer at the Lowell Observatory some three
hundred miles away in Flagstaff. The five minutes I waited
seemed like an eternity before Brian returned my call to say
"You better send a telegram. You have a comet."

Levy 159   NGC 2264
Christmas Tree open star cluster in Monoceros
First seen: November 4, 2002
Position (2000.0): α 06 41.1   δ +09 53
Magnitude: 3.9
Distance: about 2,400 light-years
Best seen: in winter; observable in city sky
Shapley class c; Trumpler class IV 3 p n

This cluster really does resemble a Christmas tree, complete
with a bright star at the top. You do not need a dark sky to

make out its beauty, since it is visible from suburban locations with the naked eye. And binoculars or a small telescope will reveal the spectacle from virtually any location.

The Christmas tree shape is best appreciated if you use a telescope, like a Newtonian reflector, that has south at the bottom. The bright variable star S Monocerotis—a huge, fast-burning O-type star—is at the base of the tree. The star at the top (or south in the inverted image) is V429 Monocerotis, a young variable star that tends to flicker over short periods of time, like the stars in the Orion Nebula. Just to the south of this star is a faint nebula that surrounds a dark, cone-shaped structure. Although the Cone Nebula appears in photographs, it is very difficult to see through a telescope.

Observing from Lake Sonoma on the cold winter evening of February 22, 2003, Jane Houston Jones, one of the foremost amateur astronomers in the United States, wrote an excellent description of the Christmas Tree cluster:

> The region around S Monocerotis is a fascinating mixture of red fluorescent hydrogen and dark, obscuring dust lanes. Some dust patches are close enough to bright stars to reflect light from them. Some of the wispy tendrils of nebulosity are Herbig-Haro objects, jets of matter ejected from newly formed stars still hidden within the nebula. At the eyepiece, we see the mag 3.9 Christmas Tree cluster with S Mon as the tree trunk. The cluster is surrounded by emission nebula and under excellent transparent skies, the elusive Cone Nebula may be visible. But not to the observers at Lake Sonoma. Some of us did, however, see some of the wispy Herbig-Haro objects visible as horizontal streaks in the nebulae below S Monocerotis.[2]

What are Herbig-Haro objects? First described in the 1950s by George H. Herbig and Guillermo Haro, "Herbig-Haro objects" are small nebulae that vary in brightness on an irregular basis. They tend to appear near the edges of dark nebulae and are thought to be either protostars or newly formed stars hidden by dust. They typically coexist with the winds of protons and electrons that stream from nearby stars.

HOW STARS ARE BORN

Since the Christmas Tree cluster is a stellar nursery, it is time to pause to explore the fascinating question of how stars are born. The process of star formation is based in matter condensing out of a cloud of gas and dust, until ignition occurs and the star is born. This process involves both stars and nebulae.

In the Christmas Tree cluster we can see how it takes place. The stars in this cluster represent a group of stellar children whose formation is essentially complete. Some of the stars are less than a million years old. Given another several million years, more of the gas and dust in the nebular regions of the cluster will coalesce to produce new stars.

In the area where stars are in the process of birth, the highlight is the Cone Nebula and its associated young star, V429, just north of the Cone Nebula. The young variable stars are members of what we call a "T association" of stars that have formed from a common cloud of dust and gas. They are an intricately woven portion of the cosmic fabric of star and nebula. Young stars vary for several reasons, one of which may be their passage through differing thicknesses of nebulosity. These stars are also believed to vary intrinsically. The result is a group of stars whose complicated patterns of behavior display two or more types of variation.

As you look at clusters like the Christmas Tree, or the more popular Great Nebula in Orion, you can imagine the incredible processes that take place, over several hundred thousand years, to result in the birth of a star. A star's birth is a very private thing, in the heart of a very thick deposit of gas and dust called a Bok globule. We'll learn more about Bok globules in chapter 6.

Levy 160   NGC 2254
Open star cluster in Monoceros—"Mountains in the Sky"
First seen: November 4, 2002
Position (2000.0): α 06 36.0   δ +07 40
Magnitude: 9.7
Distance: about 7,100 light-years
Best seen: in winter; observable in city sky
Beautiful open cluster
Shapler class f; Trumpler class I 2 p

On the night of November 4, 2002, the Christmas Tree cluster was so magnificent that I thought I would rank it as one of my favorite objects. But as I continued southward in my comet search, I came across something even better. NGC 2254 is much fainter than the Christmas Tree, and it is even fainter than its neighbor Cluster 2251, but its rich pattern of stars appear in a configuration that seems to rise and fall like a mountain range viewed in the distance on a misty day. This is NGC 2254, the cluster I call "Mountains in the Sky." Jane Houston Jones wrote about nearby NGC 2251—it and NGC 2254 "are bisected by a line of bright stars. I kept bumping into the line of stars, so it was easy to differentiate larger mag 7.3 NGC 2251 from smaller mag 9.7 NGC 2254. NGC 2254 is 2.2 degrees south of the Christmas Tree cluster. Much smaller NGC 2254 is a degree south of this."[3]

Levy 198   NGC 2437 M46
Open star cluster in Puppis
First seen: March 15, 1983
Position (2000.0): α 07 41.8   δ −14 49
Magnitude: 6.1

Distance: about 5,400 light-years
Best seen: in winter; observable in city sky
Beautiful open cluster with planetary nebula NGC 2438
Shapley class f; Trumpler class III 2 m

One of my favorite Messier objects, M46 is unique in that it is a bright open star cluster with a planetary nebula that seems to be a part of it. M46 is a circular conglomeration of stars. However, the nebula is "only" about 2,900 light-years away, about half the distance of the cluster, so it is by no means a member of the cluster.

Levy 254   NGC 869 and 884, the double cluster in Perseus
Magnificent pair of open clusters
First seen: October 28, 1962
Position (2000.0): α 02 19.0 δ +57 09
Magnitude: Each cluster is 4 magnitude and a half degree in diameter
Distance: more than 7,000 light-years
Best seen: in fall and winter; observable in city sky
A magnificent pair of open clusters
869: Shapley class f; Trumpler class I 3 r
884: Shapley class e; Trumpler class I 3 r

Although I first saw the double cluster from the city sky outside the Jewish National Home for Asthmatic Children, where I was living during fall 1962, it wasn't until summer 1968 that I really appreciated this cluster for the first time. Setting up Minerva, my 6-inch f/4 reflector, with a 16.3 mm "Galoc" eyepiece, I was completely astounded by the richness of the field of view. (A six-inch mirror that brings its light to a focus twenty-four inches away has a focal ratio of 4; we say f/4.) The double cluster is simply one of the most gorgeous things in the night sky.

Levy 301   NGC 2670
Open star cluster in Vela
First seen: November 8, 2004
Position (2000.0): α 08 45.5   δ –48 47
Magnitude: 7.8
Distance: about 3,200 light-years
Best seen: Southern Hemisphere object visible from southern US lati-
    tudes
Open cluster
Shapley class d; Trumpler class II 2 p

I found this interesting open cluster while comet hunting on a
cool November morning. Morning twilight had already begun
and I was just about to stop when the cluster came into the
field of view. It is remarkable in that the stars appear in the
shape of a bow and arrow!

Levy 305   NGC 4755, The Jewel Box
Open star cluster in Crux
First seen: November 13, 2004
Position (2000.0): α 12 53.6   δ –60 20
Magnitude: 4.2
Distance: about 7,600 light-years
Best seen: strictly in the Southern Hemisphere
Bright open cluster
Shapley class g; Trumpler class I 3 r

Hanging off the Southern Cross very close to Beta Crucis, this
beautiful cluster is part of the reason that observers travel to
the Southern Hemisphere. It is quite dense with stars. Its bright
stars form a letter A with faint stars mostly on the east side of
the A. At an age of about seven million years, it is one of the
youngest known clusters.

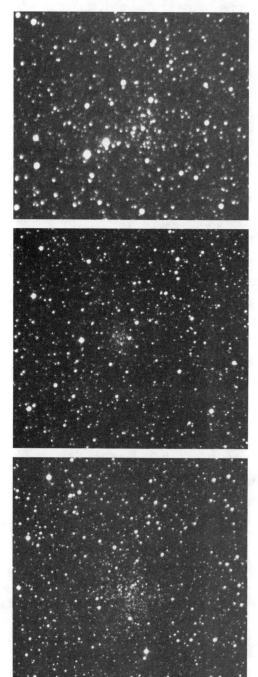

Levy 320, Tombaugh 1 cluster.

Levy 321, Tombaugh 2 cluster.

Levy 322, Tombaugh 3 cluster.

*Levy 324, Tombaugh 4 cluster
with nebula IC1795
at upper right.*

Levy 325, Tombaugh 5 cluster.

Levy 322   IC166; Tombaugh 3
Open star cluster in Camelopardalis
First seen: January 5, 2005
Position (2000.0): α 01 52.5   δ +61 50
Magnitude: 11.7
Distance: about 11,000 light-years
Best seen: in fall and winter; observable in dark sky
Faint open cluster; Dreyer thought it was nebulous
Shapley class c; Trumpler class III 1 r

After a rare Tucson rainstorm that lasted several days into the
night of January 5, 2005, the sky finally cleared late in the

evening. That's the great thing about rain and a cold front—it really gets the dust out of the air and leaves it dark and clear enough so an observer can locate some of the faintest objects in this list, like the open star cluster IC166, also called Tombaugh 3.

"Dr. Slipher, I have found your Planet X!"

During his search for planets beyond Neptune, Clyde Tombaugh became famous for his discovery of the planet Pluto. But he made other discoveries, including five open star clusters, of which this was the third.

GETTING TO KNOW CLYDE

I first met the great Tombaugh vicariously one evening during summer 1960. At dinner, Dad eloquently told the exciting narrative of the discoveries of the three outer planets, Uranus, Neptune, and Pluto, the outermost worlds of our solar system. The story inspires the mind: in one hundred fifty years, the size of our solar system increased four times. It began with William Herschel's discovery of Uranus as part of his survey of the sky, which in turn led to the mathematical prediction of the position of Neptune by John Adams and Jean Leverrier, and finally to Lowell's search for and Clyde Tombaugh's discovery of Pluto.

Two years after hearing Dad's story, I met the great observer in Denver at the 1963 Symposium on the Exploration of Mars. We were listening to rocket scientist Willey Ley's humorous introductory speech, and then the conference was thrown open to questions. Prospective questioners had to stand up, go to a microphone, and state their names and affiliations. When the first man to approach the microphone identified himself as "Clyde Tombaugh, New Mexico State," my heart skipped several beats. To be in the same room with the great discoverer was all the excitement I would ever need. But the encounter was again vicarious, since the closest we ever got to each other was about one hundred feet in same auditorium. I didn't even have a chance to approach him and say, "Dr. Tombaugh, in twenty-five years I will write your biography, and you will be partly responsible for my marriage to Wendee."

That thrill was matched at last, seventeen years later during summer 1980, when I actually got to spend time with Clyde. Between that day and his death in January 1997, we got to be good friends. On one of my frequent trips to his home in Las Cruces, I met and fell in love with Wendee, a physical education teacher. We were married in 1997.

Fourteen years earlier in 1983, after spending several nights making sure I had it right, Clyde and his wife, Patsy, came to visit, and I set up Miranda, my 16-inch reflector. It was one of life's precious moments to be able to show the discoverer his own planet. In my observing log for that evening he wrote, "Sure enjoyed looking at the many objects with your 16-inch telescope, especially Pluto, the Hercules Globular cluster, and the Whirlpool Nebula, M51. I could see the spiral structure in M51. Also, M13 was superb! But poor little Pluto was so faint, unimportant looking. . . . Thank you for this observing session. —Clyde W. Tombaugh"[4]

On January 5, 2005, I spent a chilly two hours outside reacquainting myself with the man whom I missed so much. The way I accomplished this was to search for two of his open clusters, IC166, also known as Tombaugh 3, and Tombaugh 5.

A REMARKABLE OPEN CLUSTER

I found that the cluster Tombaugh 3 was the toughest of the Tombaugh clusters to find. It is quite faint, despite the remarkable clarity of the sky that night. Its stars are mostly very faint, perhaps 15 or 16 magnitude. The brightest stars are probably superimposed on it, closer to us than the fainter members of the cluster, because they are no more thickly distributed than the stars in other areas of the field. What makes the cluster interesting is that the background of faint stars shows up so easily against the dark background; however, I did not see any evidence of the nebulosity that Dreyer reported.

Levy 324   Tombaugh 5
Open star cluster in Camelopardalis
First seen: January 5, 2005
Position (2000.0): α 03 47.8   δ +59 03
Magnitude: 8.4
Distance: about 5,900 light-years
Best seen: in fall and winter; observable in dark sky
Faint open cluster
Shapley class d; Trumpler class III 2 m

Not far from one of the brightest open clusters in the sky—the double cluster in Perseus—lies one of the faintest: a small, distant cluster proudly bearing the name Tombaugh 5. It is the brightest of Tombaugh's family of clusters. It has a few dozen faint stars in a loose, irregular group about a quarter of a degree in diameter. It has no prominent central star, but some of the brighter stars near its center seem to take the rough shape of a cross.

Levy 330   NGC 457
E.T. Open Star Cluster in Cassiopeia
First seen: January 9, 2005
Position (2000.0): α 01 19.1   δ +58 20
Magnitude: 6.4
Distance: about 10,000 light-years; Phi Cas (probably not a true member of the cluster) between 1,000 and 4,000 light-years
Best seen: in fall and winter; observable in city sky
A rich open cluster
Shapley class e; Trumpler class I 3 r

Some of the objects in this book remind me of friendship, and NGC 457, the magnificent open star cluster in Cassiopeia, is one of them. This one goes back to an astronomy shop in northwest Tucson and its owner, Dean Koenig. In addition to selling telescopes, accessories, books, and other astronomy-related items, Dean conducts a series of evening "astronomy tours," both outside the store and at various hotels, resorts, schools, and other locations around the city. I've been to many such events, enough to know that Dean's favorite object in the whole sky is a cluster around the star Phi Cassiopeiae.

Friendship is an appropriate name for this cluster, which really looks like a person with arms outstretched in friendship. For the last twenty years the cluster has been affectionately known as E.T., because of its two bright stars that resemble the eyes of the famous little extraterrestrial. I don't like its other moniker, the Owl Cluster, since there is also the Owl Nebula, Messier 97 (see next chapter). Others call it the Dragonfly Cluster, or even Caldwell 13. For Steve O'Meara, however, "This pack of glittering suns looks like the stick figure of a man ready to greet you; two prominent stars at the cluster's southeastern end mark the figure's gaping eyes; two chains of stars on the northeastern and southwestern sides of the cluster's tapered body extend outward like wide-open arms; and two isolated signs at the clusters northwestern end mark the figure's comfortably spaced feet. The cluster looks like it is not only happy to see you but wants to give you a hug."[5] The two prominent stars that are the eyes are the two components of the double star Phi Cassiopeiae. Discovered by William Herschel in 1787, the cluster lies in the Perseus arm of our galaxy, but its distance is still unknown for certain.

Levy 336   M45
Pleiades star cluster and reflection nebula
First seen: September 1, 1961
Position (2000.0): α 03 47.0 δ +24 07

Magnitude: 1.5
Distance: 400 light-years
Best seen: in fall and winter; observable in city sky
Shapley class c; Trumpler class I 3 r n

> Many a night I saw the Pleiads, rising thro' the mellow shade,
> Glitter like a swarm of fire-flies tangled in a silver braid.
> > —Alfred, Lord Tennyson, "Locksley Hall," 7–10, 1842

> I am the owner and the sphere,
> Of the seven stars and the solar year,
> Of Caesar's hand, and Plato's brain,
> Of Lord Christ's heart, and Shakespeare's strain.
> > —Ralph Waldo Emerson, "The Absorbing Soul"

In this chapter I save the best for last, the great Pleiades cluster, or the seven sisters. But why seven sisters, when there are only six members that are bright enough to be seen with the naked eye? Even Japan's Subaru (which is a translation of Pleiades) company has only six stars on its symbol. Most likely, a seventh has faded since ancient times. Through a telescope you can see many fainter stars.

I have loved this cluster ever since I noticed it as a fuzzy patch of mottled light rising in the east as I walked home on the night of September 1, 1961. I fell in love with it again when, five years later on the night of July 8, 1966, I saw it this time from the dark sky of the Adirondack Science Camp. Despite the presence of a brilliant northern light display that night, I had no trouble seeing the faint cloud surrounding Merope, one of the cluster's brightest blue stars.

On November 4, 1989, the Shoemakers and I photographed the Pleiades twice as part of our comet and asteroid search. On those two images we found asteroid 1989 VA, an "Aten" asteroid that orbits the Sun in less than two-thirds of a year, one of the fastest of any known objects in the solar system.

CHAPTER 6

CLOUDS OF DUST AND GAS
DISTANCES: THOUSANDS OF LIGHT-YEARS AWAY

To man is permitted the contemplation of the skies . . .

> —Samuel Johnson, *Rasselas*, 1759

When shall the stars be blown about the sky,
Like the sparks blown out of a smithy,
   and die?
Surely thine hour has come,
   thy great wind blows,
Far-off, most secret, and inviolet Rose?

> —William Butler Yeats,
> "The Secret Rose," 1899

Our ship sails outward, stopping by the Great Nebula in Orion. Here, as in other clouds of gas and dust about the sky, we witness the life cycle of stellar events. Whether a nebula is a place where stars are being born, as it is here in the Great Orion Nebula; or a spot where a star has expelled its outer envelope, and winds from that star are blowing that envelope about (as we shall see in a planetary nebula); or the gravesite of a star that has utterly destroyed itself in the explosion of a supernova—nebulae are places where stars seem, as Yeats wrote, to blow about the sky.

It is only some forty years ago that the term "nebula" ceased to refer to anything in the sky that is not a star. As a young member of the Royal Astronomical Society of Canada in Montreal, in the early 1960s I learned that people were still confusing the terms "nebula," which now refers to the clouds of dust and gas within our own galaxy, and "galaxy," which is the basic large gathering of stars in the universe. Back then we were still making the mistake of referring to the Andromeda Nebula, when, in fact, we should have been calling it the Andromeda Galaxy. The director of our observing program at the Royal Astronomical Society's Montreal Centre, Isabel Williamson, felt so strongly about using the correct terminology that she announced that anyone caught uttering the words Andromeda Nebula would have to put 25¢ into the observatory's piggy bank. The term extragalactic nebula, also popular at the time, has thankfully now been replaced by "galaxy." Nebulae exist in galaxies.

THE NATURE OF
BRIGHT AND DARK NEBULAE

From an observational standpoint, nebulae are best described as bright or dark. Bright nebulae are gas or dust clouds, usually in combination, that emit or reflect light. Dark nebulae are dust clouds that obscure light from behind them. Composed of opaque dust grains, they become "visible" only if they block the light from objects that are behind them. This cosmic dust consists of carbon, silicate, or iron particles about the size of cigarette-smoke particles. Each particle is coated with water, ice, methane, and ammonia. The vast majority of cosmic dust clouds in our galaxy cannot be seen.

The selection that follows begins with the nebulae associated with star birth, continues with the planetary nebulae that result as stars age, and ends with the nebula formed as a result of the total collapse of a star.

Levy 1   NGC 1931
Nebula in Auriga
First seen: January 1, 1966
Position (2000.0): α 05 31.4   δ +34 15
Magnitude: about 9
Distance: perhaps 6,000 light-years
Best seen: in fall and winter; observable in city sky

> My love, her mistress, is a gracious moon;
> She, an attending star, scarce seen a light.
> —William Shakespeare, *Love's Labour's Lost* 4.3.226–27, 1595

When Leslie Peltier began searching for comets in 1922, he also resolved to keep a catalog of the objects he called "comet masqueraders," just like Charles Messier did more than a century before. Because his telescope was more powerful than that of Messier, he knew that few of his entries would be the same as Messier's. But as I read his words I realized how exciting it would be to keep a catalog in the Messier-Peltier tradition. So on December 17, 1965, when I began my comet hunt, I wondered how long it would be before I encountered my own first entry.

Levy 1, NGC 1931.

Despite the poor weather that is typical of Montreal at that time of year, I didn't have long to wait. On New Years Day 1966, I set up Pegasus, my 8-inch reflector, to do an evening of comet hunting. My goal that night was to search for the constellation of Auriga the charioteer. I knew it was home to some glorious open clusters, the ones Charles Messier had identified as Messier 36, 37, and 38, but I did not expect to see a dim patch of haze, appearing like an out-of-focus star. A check of my atlas confirmed that it was NGC 1931, a condensed open cluster enmeshed in some nebulosity.

NGC 1931 was a great object with which to start a catalog. There is nothing obvious about it—it is small enough that it could be a faint comet, but it doesn't seem to belong with the nearby open clusters. It looks like a star out of focus through a small telescope, but through Miranda, it looks like what it is—a small star cluster. It is surrounded by nebulosity, which differentiates it from other clusters that are similar in size.

NGC 1931 is the opposite of the Pleiades, in a sense. They are both clusters of young stars with reflection-type nebulae. With the Pleiades, the hot blue stars totally dominate, and the nebulosity, except around the star Merope, is barely visible. With NGC 1931, the nebulosity is so completely dominant that it washes out the stars.

REFLECTION AND EMISSION NEBULAE

In the years since I first encountered NGC 1931 I've learned much more about nebulae. Hundreds of them are visible through small telescopes: they are softly growing clouds of hydrogen gas and pockets of grains of dust shining brightly. Nebulae are also special in that every one appears different in size and shape.

There are two basic types of bright nebulae. Emission nebulae are dark clouds of hydrogen gas glowing because their atoms have been ionized by the young, hot stars embedded within them. These nebulae shine from the energy they receive from the stars. Reflection nebulae are composed mostly of dust. We can see them only when they reflect light from nearby stars. As a consequence, reflection nebulae tend to be fainter than emission nebulae. NGC 1931 is a combination of both.

Levy 14   NGC 2068 M78
Nebula in Orion
First seen: January 9, 1965
Position (2000.0): α 05 46.7   δ +00 03

Magnitude: 8.0
Distance: about 1,600 light-years
Best seen: in winter; observable in city sky

On September 10, 1966, while comet seeking under the bright lights of Montreal, I chanced upon an object that looked so much like a comet that I really thought I had bagged my first one. But on inspecting a star atlas, I found to my disappointment that my comet was none other than Messier 78, a bright nebula in Orion north of the Great Nebula. On checking my records, I realized that I had actually spotted this nebula, on purpose, early in January 1965. What a wonderful feeling, though, actually to start searching for comets and appear to find M78 only hours later!

I wasn't the only one to feel that way. Charles Messier himself stumbled over it during his comet search more than two hundred years ago, and Leslie Peltier also noted M78 as being a major test object for the serious comet hunter. The whole idea is to be able to tell the difference between a comet and a deep sky object like M78 that masquerades as one. If what you see is a comet, it will appear to move against the background stars over the short period of time. If on the other hand it is a nebula, or a galaxy, it will not appear to move and will stay plastered against the stars forever. M78 is one of the best objects that we have to train people in the search for comets.

V351 Orionis is a variable star associated with M78. I found that by estimating its magnitude every ten or fifteen minutes over a two-hour period each night, the star would often put on a show, changing over a few tenths of a magnitude.

Levy 75   NGC 3372
Eta Carinae Nebula
First seen: March 19, 1987
Position (2000.0): α 00 47.0   δ −11 52

Magnitude: 8.0
Distance: about 1,600 light-years
Best seen: in fall; observable in suburban sky

> Now, by the sky that hangs above our heads,
> I like it well.
> —Shakespeare, *King John* 2.1.397–98, 1596

Eta Carinae is an outstanding combination of star and nebula. At low power, the nebula is complex and lovely, one of the most stunning sights in the sky. However, the reason I have included it in the catalog is personal, for it reminds me of Bart Bok.

Shortly after he arrived at Harvard in 1930, Bok began work on his dissertation about Eta Carinae, a system he thought to be unique in the galaxy. At the same time he married astronomer Priscilla Fairfield, and over the course of their lives they spent some time in the Southern Hemisphere observing, photographing, and otherwise studying this beautiful nebula. His mentor at Harvard, Harlow Shapley, teased him about his passion for the nebula, saying that his dissertation should be entitled "Miscellaneous Nonsense Vaguely Related to Eta Carinae."

Despite Shapley's jest, he knew that it was not at all nonsense! He seriously wanted Bok to study the star that first attracted attention in 1677 when Edmond Halley noticed its brightening to magnitude 4. In 1827 it swelled to magnitude 1, and a year later it had faded only half a magnitude. In 1843 it shone briefly at magnitude *minus* 1.5; thus it tied with Sirius for being *the* brightest star in the sky. In the twentieth century it has varied between magnitudes 6.5 and 7.9.

Now Bok was returning to Shapley's original suggestion for a thesis about Eta Carinae, whose "grand sweep of the swirling gases" of the nebula surrounding the star was so interesting to him. "I saw in that region an important part of the Milky Way," Bok noted; here was a good place to begin looking for examples of how young, hot stars are distributed in space. Thus the dissertation began to take form. From a study of the distribution of stars in the region of Eta Carinae, Bok was hoping to shed some light on the structure and rotation of our galaxy. He would later expand his thesis to work with problems involving the stability of clusters, their disintegration, and other questions that arose because of the galaxy's rotation. It was a project that occupied Bart and his wife throughout much of their lives.

A few days before her death in 1975, Priscilla and Bart were attending the opening of the Flandrau Planetarium in Tucson. As the couple walked past the pictures, Priscilla stopped at the one showing Eta Carinae. "Bart," she said, "when I die, this is where I want to be. I will be watching you from here." As I talked with Bart in his last years, he often mentioned Priscilla and Eta Carinae. Bart died in 1983, and I share the feeling of his family and friends that he and Priscialla are in the Eta Carina Nebula.

I first spotted Eta Carinae in 1987. I spotted it again on February 9, 2002, while comet searching with Ana Guillermina Reyes's 20-inch reflector at the Asociacion Salvadoreña de Astronomia Observatory in El Salvador. Through the eye of this mighty telescope, my comet hunting paused abruptly when the nebula swam into the field of view. That was the night I hunted along the most gorgeous strip of the entire sky: from Alpha and Beta Centauri, through the Coal Sack and the Southern Cross, and ending with Eta Carinae.

I saw Eta Carinae again with Wendee and a group of other observers at an ancient Aboriginal rock carving site south of Alice Springs, Australia, on the night of the Leonid meteor storm of November 18/19, 2001. As Wendee looked through the 3.5-inch telescope at the nebula, I felt that she was beginning to meet Bart Bok and appreciate all that he accomplished.

Levy 76   NGC 6618 M17
The Loon, the Swan, the Horseshoe, or the Omega Nebula
First seen: July 26, 1965
Position (2000.0): α 18 20.7   δ −16 10
Magnitude: 6.0
Distance: 4,890 light-years
Best seen: in summer; observable in city sky

MY FAVORITE MESSIER OBJECT

To determine which of all Charles Messier's 110 objects was my favorite, I tried to see them all on one of the few nights of the year that they all are visible—in the middle of March. On March 14, 1983, I set out to see them all, but clouds rushing in the last hour before dawn forced me to stop with ninety-one Messiers. The following night's sky was very clear. I managed to find every Messier object except Messier 30 in Capricornus, which rose after twilight and had become too far advanced. In my observing log for that night, I wrote for M17: "Running out of words to describe these magnificent objects. The nebular filter provides one of the finest sights—Earth or sky—that I have ever seen." (On that memorable night, M17 beat out 108 other Messier objects and 34 other members of the NGC .) Philippe Loys de Chéseaux discovered it in 1746, two years after the appearance of his great comet.

Nebulae like this one are similar to Rorschach inkblot tests; they can mean different things to different people. To many, it is a checkmark, its long arm stretching through 12 light-years. George Chambers called it a swan in 1889. Roy Bishop of Avonport, Nova Scotia, says it resembles a loon more than a swan. And the famous French astronomy writer Camille Flammarion said it resembles a "smoke-drift, fantastically wreathed by the wind."[1]

Levy 76, M17. Photograph by Tim Hunter. All Tim's pictures were taken with a Meade LX200 12-inch f/6.5 telescope and an Apogee AP 7 CCD camera.

Levy 77   NGC 2237, 2238, 2244
Rosette Nebula; Swift's Nebula
Position (2000.0): α 06 32.3   δ +05 03
Magnitude: 6.0
Distance: 4,900 light-years
Best seen: in winter; observable in dark sky

Much larger than the Orion Nebula (its wisps of nebulosity are more than a hundred light-years wide) but farther away, the combination of the Rosette Nebula and its associated cluster of stars are best viewed under a dark sky. The elegant strands of bright nebula are interlaced with dark veins, and inside the veins are "Bok globules."

The nebula was independently discovered by Lewis Swift in the early 1880s, during his search for comets. In 1883 Edward Emerson Barnard also found it as part of *his* comet search. The nineteenth century is often called the golden age of comet seeking, when Lewis Swift, Horace Tuttle, and Edward Emerson Barnard were competing for new comets.[2]

It wasn't just comets that these people found. Their efforts uncovered beautiful, and permanent, sights in the sky like the Rosette Nebula and its dazzling cluster of young stars.

Levy 78   NGC 2261
Hubble's Variable Nebula
Position (2000.0): α 06 39.2   δ +08 44
Magnitude: about 10

*Levy 77, The Rosette. Photograph by Dean Koenig using a
4-inch Astrophysics refractor and SBIG ST-10 CCD camera.*

Distance: 2,500 light-years
Best seen: in winter; needs dark sky

Even though William Herschel discovered this nebula in
1783, it was Edwin Hubble who made it famous. In fact, the
nebula is so often associated with Hubble that people who
enjoy it have long forgotten that it was the great Herschel who
first saw it. To remind us, Steve O'Meara suggests that we call it
"Herschel's forgotten fan."[3]

In 1914 a young student named Edwin Hubble joined the
staff of Yerkes Observatory and began an investigation of the
nebulae (as the old term, all nonstellar objects). As a junior
astronomer at the observatory, Hubble did not often use the

observatory's gem of a telescope, its 40-inch refractor. But he did make good use of the smaller 24-inch reflector there. It wasn't long before Hubble made his first discovery at Yerkes. He decided to try photographing a strange nebula, NGC 2261, in the constellation of Monoceros, that looked like a comet but which we now know to be a nebula that is lit by the young variable star R Monocerotis. The star's light both reflects off the nebula back into space and shines through it, causing the nebula's particles to ionize and emit it again; thus the nebula is both a reflection and an emission nebula. Hubble photographed it many times during the observing season that began in fall 1915. When he compared his pictures to images taken in 1908, he saw that the nebula's west side had expanded and had become more radically convex. He showed his images to Barnard, who was then also at Yerkes. Taking advantage of Barnard's long observing experience, Hubble worked with him to confirm his discovery. It would have been extraordinary to witness these two men working together: Barnard, who had put his mark into the sands of time by sheer hard work, and Hubble, who was poised in that year to become the astronomer who would redefine our place in the Universe.

The nebula whose variation Hubble discovered is associated with a variable star called R Monocerotis; it varies irregularly by about half a magnitude around 11. However, the star is usually very hard to see because it is embedded in the nebula. The variation does not seem to follow the brightness changes in R Monocerotis, and they do not occur with any regularity.

R Monocerotis and its nebula probably represent a planetary system in an early stage of formation. At least two other variable nebulae are known. One is NGC 1555 in Taurus with a tiny wisp in Corona Austrina; the other is NGC 6729, the home of R Coronae Austrinae.

Levy 95   NGC 1999
Bright nebula in Orion
First seen: November 15, 1979
Position (2000.0): α 05 36.   δ –06 42
Magnitude: 10.5
Distance: about 1,500 light-years
Best seen: in winter; observable in city sky

Surrounding nebular variable star V380 Orionis, this reflection nebula was pretty much ignored because it lies so close to its more famous cousin, the Great Orion Nebula. However, in 1999 the Hubble Space Telescope took a phenomenal image of the nebula. I have encountered this always-welcome sight several times during comet hunting.

Levy 112   NGC 7023
Reflection nebula and star cluster in Cepheus
First seen: August 13, 2002
Position (2000.0): α 21 00.5   δ +68 10
Magnitude: 7.7
Distance: about 1,400 light-years
Best seen: in spring; observable in suburban sky

This is a most unusual-appearing reflection nebula around the star cluster Collinder 429. The star is close to the star Beta Cephei, by only three degrees, and it has a bright star in its center.

Steve O'Meara points out that this nebula is just one degree, or two moon-diameters, from one of my favorite variable stars, T Cephei. It varies from magnitude 6 to 10 in a little more than a year, actually 387 days. The American Association of Variable Star Observers Web site has a series of charts for T Cephei; they are for the taking, but do consider joining this worthy organization. Variable star observing is a fascinating field, and T Cephei is one of the easiest stars to observe.

Levy 126   NGC 2359 Thor's helmet
Bright nebula in Canis Minor
First seen: October 1, 2002
Position (2000.0): α 07 18.6   δ −13 12
Magnitude: about 10
Distance: at least 10,000 light-years
Best seen: in winter; requires dark sky

> [The knight errant] must be an astrologer, so as to tell by the
> stars how many hours of the night have passed, and what
> part of the world he is in.
> —Miguel de Cervantes Saavedra, *Don Quixote* 2, 1615

Also called the Duck Nebula, this strange object looks like a
ghostly Viking helmet in the sky—at least in photographs. I've
encountered this object many times during comet hunting.
This patch of haze is easier to see with a UHC or nebular filter.
The nebula is associated with a Wolf-Rayet star called HD (for
Henry Draper's Catalog) 56925. This Wolf-Rayet star is very hot
(from 25,000–50,000 K) and huge (some twenty times the mass
of the Sun). As winds of material blow off the star at more than
4 million miles per hour, it loses mass into a gas bubble that we
see in the shape of a helmet. This star is undergoing a brief
phase of instability by which it loses mass quickly. Once it has
lost some of its mass, it will become more stable again.

Levy 150   NGC 6523 M8
Bright nebula in Sagittarius
First seen: August 12, 1963
Position (2000.0): α 18 03.8   δ −24 23
Magnitude: 3.0
Distance: 5,200 light-years
Best seen: in summer and fall; observable in city sky

Levy 150, M8.

When John Flamsteed discovered the Lagoon Nebula in 1680, he had no idea that it would become as prominent an object for study as it has. On May 23, 1764, Messier found it independently and added it to his catalog. He recorded that both a cluster and a nebula were involved here. Steve O'Meara is especially poetic in his description of the Lagoon: "Particularly delicate at 23x, the nebulosity and its myriad dark lanes look like a frozen flower petal that has fallen to the ground and shattered. If you mentally erase the nebulosity, you should see a crossbow of seven prominent stars—the skeleton of this Messier object."[4]

O'Meara also recommends that you spend plenty of time studying the Lagoon's center. Its main features are knotty bright material and a wishbone of dark lanes to the northwest. These bright knots make up the Hourglass Nebula.

BOK GLOBULES IN THE LAGOON

Deep in the Lagoon Nebula are some small, dark nebulae called Bok Globules, which were mentioned earlier. They were first cataloged as dark nebulae by Edward Emerson Barnard. Then at Harvard just after the end of World War II, Bart Bok took a special interest in these dark nebulae. Around 1947 Edith Reilly, a young associate, asked Bok if she could study dark nebulae with him. With his wartime navigation work over, Bok was eager to spend more time on his Milky Way research, and the dark areas were certainly high on his wish list. He had especially hoped to develop a classification scheme and was looking for someone to investigate the nebulae on Harvard's plates. Unfortunately, Reilly, with multiple sclerosis, was not physically strong enough to carry the heavy 8-by-10-inch photographic plates and place them under microscopes for observation.

Bok soon saw a perfect opportunity in Reilly's request. Why not have her examine the old catalogs of several hundred dark nebulae that Barnard had prepared, to determine which of them would be candidates for further study? At first the work was relatively routine, but when she progressed through Barnard's list, Reilly found notes about very small, round, and unusually dense nebulae. His interest aroused, Bok began photographing these nebulae. Forgetting about their classification project, Reilly concentrated on identifying these nebulae while Bok set out to photograph them using the Jewitt Schmidt telescope at Harvard's Oak Ridge Observatory station. In 1947 a preliminary paper discussed their work on these "small dark nebulae."[5] They were typically round, from three to five arcminutes wide (about one-sixth the Moon's diameter of the full Moon). "Through a telescope," Bok later described, "you would come to the leading edge of one of these things and suddenly the stars would just disappear. And then you would push the telescope's slow motion button a bit and bloop! The stars come back."

Bok and Reilly found about two hundred of these dark objects within the relatively close distance of 1,500 light-years. These tiny nebulae were optically extremely thick, with possibly *thirty magnitudes* of extinction; if one could, for example, cover a 1 magnitude star with one of these clouds, it would become invisible even through the Hubble Space Telescope.

Bok reasoned that these nebulae marked the birthplaces of new stars.[6] As their dark gases move about slowly, he reasoned, they begin a slow collapse under their own gravity that intensifies until stellar fusion starts. With evidence of their importance, Bok considered giving them a simple name. That thought was in his mind one morning as he walked downstairs, opened the front door of his Belmont home, and found bottles of unhomogenized milk delivered fresh from the H. B. Hood Company. At the top of each bottle the cream separated out, and in the cream floated small globules of fat—looking, Bok noted with astonishment, "just like my *globules!*" That's how they came to be known as globules.

What would it be like inside a globule? A large amount of unlit gas—perhaps the equivalent of nine times the mass of the Sun—would pass through the globule during its 100-million-year lifetime (which is relatively short by the standards of the Universe), and some of this gas would be captured by the globule's own frigid particles. Over this time the whole globule would shrink. If the globule were small, it would form one or two stars; if it were larger, it might form a cluster of stars.

It was when the subject of globules turned to Messier 8, the Lagoon Nebula, that some serious skepticism arose. Walter Baade from Palomar wrote to Bok in March of 1947, "I cannot convince myself that these objects are in any way peculiar. . . . I am afraid you have been misled either by the Lick print of M8 on which the objects in question may appear as inky spots on account of some stepping up process [artificially increasing the contrast on the plate] or by some Harvard plates of insufficient scale."[7] Before replying, Bok examined as many other plates of Messier 8 as he could find, including one from the large 82-inch reflector at McDonald Observatory. "From the material available to me," Bok argued correctly, "there is little doubt about the reality. . . . Edith Reilly and I have examined a number of the objects in Barnard's famous list of dark objects. As examples of objects I would be tempted to classify as globules I list the following: Barnard 34 and Barnard 92 are examples of the somewhat larger variety. They appear to be clearly defined, they are probably all relatively nearby and to my mind there can be little doubt that they are really unlit dark nebulae."[8] But Baade was unconvinced, as was his colleague Edwin Hubble. They thought instead, "that they are parts of outlying streamers of Messier 8 which are not hit by the radiation of the exciting star. The whole behavior of the streamer system suggests this interpretation."[9]

In 1956 a search of two prints of the newly completed Palomar Sky Survey revealed seventeen thousand new dark objects, averaging about one minute of arc (about a thirtieth the apparent diameter of the full Moon) in diameter.[10] When radio telescopes began studying these objects around the same time, the nature of these small dark nebulae as star precursors became much more credible.

Levy 151   NGC 6611 M16
The Eagle Nebula
First seen: July 8, 1965
Position (2000.0): α 18 18.8   δ −13 47
Magnitude: about 6
Distance: 9,000 light-years
Best seen: in summer; best in dark sky

Levy 151, M16.

Tranquillity, peaceful surroundings, the pleasures of the country-
side, the serenity of the skies . . . these are the things that en-
courage even the most barren muses to become fertile and bring
forth a progeny to fill the world with wonder and delight.
　　　　　　　　　—Cervantes, *Don Quixote* 1, 1605

It was just another beautiful Messier object I recorded, on a
warm July evening at the Adirondack Science Camp. Just how
beautiful it would turn out to be, however, I had no idea.
Exactly thirty years later, a team led by Arizona State University's Jeff Hester, using the Hubble Space Telescope, took one of
the truly great astronomical photographs of our time—the
Eagle Nebula's Pillars of Darkness.

　　Near the center of the Eagle Nebula, one of the Milky Way's
great star-forming regions, they found a specific kind of globule
that they called an evaporating gaseous globule, or EGG. It is a
small, dark region filled with evaporating gas. Bok Globules are
typically less than a third of a light-year in diameter while
EGGs are hardly more like a light-week across. An EGG, it
seems, is the womb of a single star. A longtime supporter of the

idea of a space telescope, Bart Bok would have loved that picture of a globule taken through it.

Levy 152   NGC 6514 M20
Bright nebula in Sagittarius
First seen: July 27, 1965
Position (2000.0): α 18 02.3   δ −23 02
Magnitude: 6.3, at least for the cluster associated with the nebula
Distance: 9,000 light-years
Best seen: in summer and fall; observable in city sky

Just north and west of Messier 8, the Trifid is different from its larger neighbor. Smaller and more compact, its three parts are separated by lines of dark nebulosity. The Trifid has a stunning appearance even through Echo, the 3.5-inch reflector I used to observe this object from the dark sky of the Adirondack Science Camp in summer 1965. Through a larger telescope, and especially with a nebular or UHC filter, the subtle red coloring of the main, trifid portion becomes more striking, as does the even more subtle blue of the nebulosity toward the east.

Levy 316   NGC 1976 M42
Great Nebula in Orion
First seen: September 9, 1962; added to catalog after photograph taken
    December 17, 2004
Position (2000.0): α 05 35.3.   δ −05 23
Magnitude: 3.5
Distance: about 1,500 light-years
Best seen: in winter; observable in city sky

And huge Orion, that doth tempests still portend.
          —Edmund Spenser, *The Faerie Queene* 4.9.13, 1596

Levy 152, M20.

. . . of Orion with belt and sextuple sun theta and nebula in
which 100 of our solar systems could be contained . . .
—James Joyce, *Ulysses*, 1922

A COSMIC REALITY SHOW

On the advice of J. B. Sidgwick, early in September 1962, I took
out Echo, my first telescope, to see the Orion Nebula, then in
the southeastern sky. Sidgwick's description of the nebula was
so vivid that I had to pass up sleep and get up early one
morning to look at it. "The great Orion Nebula," Sidgwick
writes, "In large instruments an incomparably grand object—a
gigantic, convoluted cloud of incandescent gas. Long exposure
photography has extended its ramifications over a great part of
the constellation. It is plainly visible to the naked eye as a
misty spot, and even in binoculars is an unusual object, a
vague mist of pale green light."[11]

When Sidgwick wrote that the manifestations of the nebula
can be detected over most of Orion, he was right. In most constellations the stars appear close to each other because of their
chance positions in the sky relative to Earth. Most of the stars
in Orion, however, are physically part of the same grouping.

Levy 316, M42.

Except for Betelgeuse, most of the bright stars in Orion are roughly the same distance—about 1,500 light-years—from us. Orion is far more than just a place in the sky that was named for a figure from mythology. Orion is a cosmic production plant, whose different divisions show us how stars *are made*. It is even related to Barnard's loop, a very large, long, curved structure to the northeast of the main body of Orion.

HOW YOUNG STARS INTERACT WITH THE NEBULA

My 1962 predawn look at the Great Nebula was the start of a fascination with this object that continues to this day. It was also the beginning of a tradition of early morning observing sessions that I enjoy so much. It's hard for me to imagine that that morning, all those years ago, was one of the first times that I set the alarm on all early morning observing sessions; predawn experiences with the stars have been commonplace in my life for many years. That night was also the beginning of the most intensive observing study of any astronomical object I have ever done. For three years, between 1978 and 1981, I spent almost all my observing hours studying the behavior of the stars within the nebula. I was watching a cosmic reality show, being a part of a family of stars, watching changes and moods, and wondering what the next night's episode would bring. It wasn't done by watching TV; it was accomplished by looking through a telescope. Over that time I made over ten thousand observations of their changing magnitudes. If someone needed to correspond with me in those years, he might as well have addressed the letter "in care of the Great Orion Nebula."

The young stars in the Orion Nebula are members of what we call a "T association" of stars that have formed from a common bond of dust and gas. They are an intricately woven portion of the cosmic fabric of the Orion Nebula, and they can vary relatively quickly. The nebula is a fishbowl through which we can see the process of star formation. Young stars in the belt represent a group of stars whose formation is essentially complete. The sword of Orion, on the other hand, is a cosmic nursery with some of its stars being less than a million years old, near infants by galactic standards. Some astronomers suspect that the area south of the nebula, with its hydrogen-rich regions, will in another several million years become transformed into new stars.

Just how that transformation takes place is still shrouded in mystery as well as the clouds of nebulosity. First there is nebulosity, and then we see young protostars, and then variable stars. We know that somehow the nebula starts contracting, but how and why? Does a nearby supernova set off the process? "Our level of knowledge of these earliest phases of contraction," writes astronomer Robert O'Dell, "is at the level of 'the stork brings them.'"[12]

After the star is born and becomes visible, there is much to see and learn. In 1939 Bengt Strömgren independently built on an idea of Bok's from almost a decade earlier that dealt with the spheres of hydrogen gas around hot, young stars. In 1930 Bok introduced the idea in a never-published paper that took advantage of the discovery that very hot dwarf stars were at the centers of planetary nebulae. The hot star at the center of nebula, he thought, releases ultraviolet radiation that ionizes the hydrogen in the cloud, causing it to glow. However, if the cloud is thin enough, some of the UV radiation leaks out through the spherical cloud. At the outer edge of the nebula, the ionized hydrogen begins to lose its ionization and becomes ordinary hydrogen. The edges of a thin nebula, where the UV is leaking out, might seem sharp. Bok explained the shells to be the sharp edges of what was later called Strömgren spheres, or still later the Hydrogen II regions surrounding stars. Since Bok's paper was never published, it was left up to Strömgren to describe these spherical regions of ionized hydrogen.[13] Strömgren spheres are zones of electrons and ions formed around a hot star embedded in a cloud of neutral hydrogen atoms. The hotter the star, the larger the H-II region around it. In this simple idea is the difference between emission nebulae, like the Orion Nebula, and other bright nebulae, like the Merope in the Pleiades, that shine merely by reflected light from nearby stars.[14]

OBSERVING THE VARIABLES IN THE ORION NEBULA

With all this exciting astrophysics going on in the sky, it's amazing that we can actually see some of it in action through a small telescope in our own backyards. Some of the hot stars in the Orion Nebula actually vary in brightness, and their behavior is worth watching! During the three years I studied them so intensely, I found that the most lively form of Orion's star behavior is a rapid and irregular "flicker" by as much as a fifth of a magnitude in five or ten minutes. I have seen V361 Orionis change brightness in this fashion. By comparing V361 to nearby stars of known, and unchanging, magnitudes, I can estimate its changes. V361 is the brightest of the rapidly changing members of the Orion family. I followed it with Pegasus, the 8-inch. I have occasionally witnessed HU Orionis (for which I need to use Miranda, the 16-inch) brighten by a fifth of a magnitude. I have especially enjoyed LP Orionis, a real troublesome child of a star. I have seen it change rapidly, but since it is embedded in a particularly dense part of the nebula, it's not easy to estimate its brightness relative to the nearby comparison stars. Faint MR Orionis was often performing, though I needed to use the larger telescope Miranda to follow its faint ramblings. It's a particularly interesting star because it is so close to the center of the nebula. For consistency, I always had to estimate this star's brightness using the same telescope night after night, with the same eyepiece. That advice is worth following for any variable star, but especially for one so embedded in nebulosity.

On the other hand, other Orion variable stars don't do very much. KS Orionis and MX Orionis never changed appreciably during my vigil. NU was more of a disappointment to me; it was advertised in the *General Catalog of Variable Stars* for having active variations, but like three Cheshire cats, NU, KS, and MX stayed still all the time I watched them.

The rapidly flickering stars vary because of some astrophysical adjustment taking place within them. But T Orionis, also in the midst of the nebula, is different. It stays at its maximum brightness for some time, and then it unpredictably drops to a fainter magnitude. T Orionis probably doesn't vary by itself; we see it fade when a thicker amount of nebulosity passes by.

Why bother observing these stars visually, especially when they can be followed more accurately with electronic detectors? There's one simple reason: you get to watch young stars at play, a real-time movie. The nebula represents an important phase in the early evolution of stars. I recommend that if you decide to observe the stars, just have fun with them; don't worry about reporting all your results, especially at first. When you do try to determine small differences in brightness, always use the same set of comparison stars.

THE CLOUDS

I was not disappointed by my 1962 introduction to the Great Nebula, and I still love it. Here is a combination of star cluster and nebulosity that simply staggers the imagination. At the center is the Trapezium, a quadruple star. How many times I've heard people looking at triple star say that they had never seen a multiple system before, forgetting that they are in all likelihood familiar with the famous Trapezium! With a larger telescope (like a 16-inch) it is possible to see two other stars, making it a "sextezium," as James Joyce noted in *Ulysses*.[15]

For all the fantastic photographs that have ever been taken of it, Wendee and I both prefer to see it visually rather than in a photograph. Wendee is no longer impressed with most of the Orion Nebula photos that appear every winter; to her they look commonplace and almost boring. However, its visual appearance is always an "Oh wow!" moment for her.

There is not one Orion Nebula; there are many, depending on the telescope we use. Through Pegasus, the subtle blues and reds are so soft that they send a chill. But through the larger aperture of Miranda, the colors are clear and distinct.

The first person to notice this remarkable cloud was Nicholas Peirac, a French lawyer, late in 1610 or early in 1611. Johann Baptist Cysatus published his own discovery notes in 1618. It was then almost forgotten until Christiaan Huygens observed it again in 1656, the same *annus mirabilis* during which he discovered Saturn's ring system.

Levy 325   IC1795
North Bear Nebula in Cepheus, also called Running Dog Nebula
First seen: January 8, 2005
Position (2000.0): α 02 26.5   δ +62 04
Magnitude: ?
Distance: about 6,000 light-years
Best seen: in winter; needs a dark sky

I first encountered this extraordinary nebula while trying to photograph L323, the fourth open cluster in Clyde Tombaugh's series. The nebulosity is clearly visible in the low-power eyepiece, but it really comes out well photographically. Long-exposure photographs give it the unmistakable resemblance to the head of an animal, like that of a bear or a fish.

Sid Leach spent several hours exposing this photograph of the complex of nebulosity surrounding Barnard 283 (Levy 356) from the shore of Lake Titicaca, Bolivia. I found the nebula while comet hunting from the same spot, and it is one of the darkest I have ever seen.

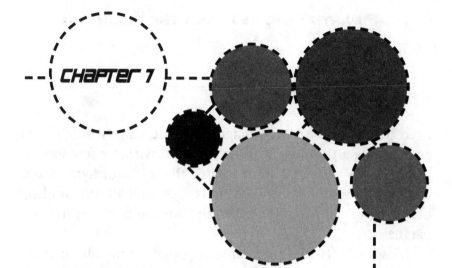

NEBULAE AT THE END OF A STAR'S LIFE
DISTANCES: THOUSANDS OF LIGHT-YEARS AWAY

[W]ho would fardels bear,
To grunt and sweat under a weary life
But that the dread of something after death,
The undiscovr'd country from whose bourn
No traveler returns, puzzles the will
And makes us rather bear those ills we have
Than fly to others that we know not of?

—William Shakespeare,
Hamlet 3.1.76–81, 1603

Levy 13   NGC 6720 M57
Ring Nebula in Lyra
First seen: July 15, 1964
Position (2000.0): α 18 53.6   δ +33 02

123

Magnitude: about 9
Distance: perhaps 2,000 light-years
Best seen: in summer and fall

Even though it is a faint 9 magnitude, the Ring Nebula is such an easy target that it is possible to find it within a few seconds through a telescope. It is located in the midst of Lyra, one of the smallest and most distinctive constellations, midway between Beta and Gamma Lyrae, the bottom two stars in Lyra's parallelogram.

Messier 57 is the most famous planetary nebula. It is the result of a star that has used its hydrogen and helium, and, in a burst of stellar temper, exploded off its outermost layers. These layers appear as a smoke ring because of the angle at which we see it. We actually are looking down from the top, through a tunnel surrounded by gas. This nebula is one of the rare deep sky objects that is beautiful through any telescope: from a small refractor that shows a puff of a smoke ring to the mighty Hubble Space Telescope (HST) that peers deeply down into the tunnel-like tube surrounding the hot, blue central star.

In the HST image, we really see straight into the tunnel. The whole nebula is only about a light-year in diameter. At its center is a small star that used to be similar to our Sun. After

spending billions of years as yellow, that star evolved to become a red giant some one hundred times larger than its earlier size. At the end of its red giant phase, it blew off its outermost layers into space. What's left is the star's core. As intense

Levy 13, M57.
Photograph by Tim Hunter.

ultraviolet radiation from the central core streams out into the surrounding gas, the gas glows.

WHAT'S A PLANETARY NEBULA?

Not all nebulae are the birthplaces of stars; as a star gets older, it can get back into the "nebular habit." An older star can shed its outer atmosphere into space; as a result, a small nebula forms and surrounds it. In summer 1764 Charles Messier discovered the Dumbbell Nebula, but the most famous one, the Ring, was found by Antoine Darquier in 1779. Darquier remarked on its resemblance to a fading planet. As the numbers of this type of nebulae grew, their vague observational connection to fading planets became apparent, and the name planetary nebulae stuck. But they are not planets; they are temporary structures of gas. Many stars are capable of producing these clouds; it is estimated that 95 percent of the stars in our galaxy will someday have planetary nebulae. The only stars that won't are the very massive ones that will become supernovae instead. The reason we don't see planetary nebulae near every star is that the stage is very short lived, lasting between ten thousand and twenty-five thousand years—less than an instant on the great timescale of our galaxy.

Our Sun will probably sport a planetary nebula someday. Right now, the Sun is a peaceful G2 star, but after some 5 billion years, it will begin to change radically and enter new phases in relatively rapid succession. In one of the later phases it will become a red giant, and then it could become a carbon star just before it sheds off its outer layers to form a planetary nebula. The thought that our own Sun, bright and warm as it is, could somehow have a future similar to the Ring makes that nebula all the more interesting to observe.

Levy 18 NGC 2392
Clownface or Eskimo Nebula in Gemini
First seen: May 5, 1967
Position (2000.0): α 07 29.2 δ +20 55
Magnitude: 9.2
Distance: about 4,000 light-years
Best seen: in winter; observable in city sky

> No natural exhalation in the sky,
> No scope of nature, no distemper'd day,
> No common wind, no customed event,
> But they will pluck away his natural cause
> And call them meteors, prodigies, and signs,
> Abortives, presages, and tongues of heaven,
> Plainly denouncing vengeance upon John.
> —Shakespeare, *King John* 3.4.153–59, 1596

If I had known that through a large telescope NGC 2392 looked like the face of a clown, it might have cheered me up more than it did, for I met that planetary nebula on a warm May night in 1967 during one of the deepest crises in my life up to that time. Just after my first full year of comet hunting, I was spending a lot of time at the Montreal Centre of the Royal Astronomical Society of Canada, a highly satisfying experience, at least until April 15, 1967, when I found that the center's wind-up barograph wasn't recording air pressure on its sheet of paper. The following night I mentioned it, an act that almost cost me my membership there and my interest in astronomy. I was blamed for overwinding the barograph. My denials only made matters worse, and within thirty minutes I was ordered to leave the building.

Just three weeks later, I was at my grandfather's country home at Jarnac, in the Gatineau hills northwest of Montreal. The magnificent dark sky there allowed me to do some quality comet searching. With each southward shift of my telescope, new fields of stars came into my view. I was searching in a rich Milky Way region of Gemini, near the place where Clyde Tombaugh discovered Pluto thirty-seven years earlier. As I looked through the telescope, my mind reverted to more peaceful times. I had recently read in *Starlight Nights*, comet hunter Leslie Peltier's exquisite autobiography, how comet hunting is a "magic carpet that can take you to nostalgic flights into the past. . . . Even now, when I am hunting late at night, there is little to remind me of what century it is. In the dark silence of the dome 200 years can disappear in just the twinkling of a thought."[1] That evening I decided that I enjoyed the night sky too much to let that incident get to me.

Just as I was thinking these good thoughts, I came across NGC 2392. Through the low power of my telescope, it looked like a faint but distinct out-of-focus star. It would be years before I'd see its detailed appearance through the 61-inch telescope at Mt. Bigelow, Arizona, as a complex oval structure that resembled either the face of a clown or of an Eskimo. At the center is a star surrounded by an oval concoction of gaseous streamers, all of which is in turn surrounded by a more circular structure that looks like an Eskimo hood. To Steve O'Meara's practiced eye, it is the "Lion Nebula"—because he sees the face of a lion peering at him.[2] Steve is right; it also looks like a lion's face.

Also, back in 1967, I knew that planetary nebulae were formed from old stars—that had been known since the 1920s. But I had no idea that ten years later a team led by astronomer Sun Kwok would propose that planetary nebulae are formed when a big red giant star, expelling its outer atmosphere, exposes the star's white core. As high-velocity winds rush out from the core, they affect the old material from the red giant, creating the strange shapes like the clown face.

I also had no idea that night that years later I'd be able to make this nebula wink at me, as if it had some sort of secret. By concentrating on the bright central star, the nebula disappears! This is a striking example of psychovisual effects on observing.

But lack of knowledge about the nature and apparent visual behavior of planetary nebulae did not prevent me from really appreciating that first small view of NGC 2392. The following night there was the most magnificent display of northern lights: just a moderate glow at 10:50 PM, but at midnight, rays appeared in the north, and then at 1:39 in the morning, they looked on fire, as shimmering arcs appeared like flames. After that weekend, I returned to Montreal full of enthusiasm for planetary nebulae, comet hunting, and the night sky with its magnificent auroral displays.

That enthusiasm didn't last long. I walked into the observatory, all excited to tell them of my observing experience, and was promptly ordered to leave. I came back but was ordered to leave once again. When I resisted this time, three people took the chair on which I was sitting and pushed it over. My NGC 2392 story still untold, I left the building. The following day, a motion was put forward at the center's board of directors meeting to expel me from the Royal Astronomical Society of Canada. If it weren't for friends like Constantine Papacosmas, and the center president who refused to allow the motion to come to a vote, I would probably not be a member today and it is possible that I would no longer have an interest in astronomy. In fact, two years after the incident, after I had moved to Nova Scotia to attend Acadia University, one of the newer members was told that "Levy will never amount to anything." Is it possible that the society was trying to demonstrate, in human terms, the idea that this nebula is the outer shell of a star that has been expelled from the system, like I would be?

I am still a member of the Montreal Centre, and ironically, their honorary president. The organization is very active in conducting star parties, visiting schools, and encouraging observing in every form.

Levy 61   NGC 246
The Cetus Ring; also Caldwell 56
First seen: June 18, 2002
Position (2000.0): α 00 47.0   δ −11 52
Magnitude: 8.0
Distance: about 1,600 light-years
Best seen: in fall; suburban sky

This object is one of the prettiest things that ever found its way into my telescope. It is a ring-shaped body that has a number of stars either enmeshed within it or just outside it. The nebula's central star and one other nearby star give it a Pac-Man appearance to some observers, but not being a game type of person, that moniker doesn't mean too much to me. It is No. 56 in the Caldwell catalog kept by the famous British amateur astronomer Patrick Moore. What caught my attention when it turned up during my predawn observing session on June 18, 2002, was its appearance as a nebula inside a constellation that is better known for faint galaxies.

If computer game aficionados want to call it the "Pac-Man" nebula, then let me call it "politics man," or "Dief the Chief" since June 18 is the fortieth anniversary of the first Canadian election whose results I followed closely. (It was the last election won by Canadian prime minister John Diefenbaker.)

Levy 88   NGC 3587 M97
The Owl Nebula
First seen: July 13, 1966
Position (2000.0): α 06 39.2   δ +08 44
Magnitude: about 10
Distance: 1,630 light-years
Best seen: in spring; suburban sky

When people refer to the Owl at star parties, I think either of a huge great horned owl that often haunts my observing nights or the Owl Nebula Messier 97. But these days they are probably talking about NGC 457, the Owl cluster. Let's call the cluster something else, for Messier 97, with its bright central star, really looks like an owl. I first saw the Owl during summer 1966 at the Adirondack Science Camp, one of the most enjoyable summers I've ever experienced. It was strange to hear the calls of nearby owls in the forest, the same summer I was looking at one in space!

Although the Owl Nebula is in Messier's catalog, it was discovered by his rival and colleague Pierre Méchain in 1781, the same year that William Herschel discovered Uranus. Messier's search for comets was the first successful one. As the great comet hunter found more comets, others joined the race. And then it became competitive. Méchain found his first in 1772, a dozen years after Messier's first, and Jacques Montaigne was finding comets by 1781.

There is a well-known but unverified story from Messier's Russian friend Frederick La Harpe. If it is true, then Messier found out about a discovery by Montaigne while Messier was mourning his wife's death. A friend embraced the grief-stricken man to say, "I am so sorry." Messier glared at his visitor. "Alas," he said, "Montaigne has robbed me of my comet!" Quickly realizing his faux pas, Messier tried to recover. "Poor woman," he muttered. No doubt his friend agreed.[3]

Levy 122   NGC 6826
Blinking Planetary Nebula
First seen: August 29, 2002
Position (2000.0): α 19 44.8   δ +50 31
Magnitude: 8.5
Distance: about 2,200 light-years
Best seen: in summer; observable in city sky

There's husbandry in heaven,
Their candles are all out.
>—Shakespeare, *Macbeth* 2.1.4–5, 1606

The Blinking Planetary Nebula, like the Eskimo Nebula, is a fine example of psychovisual effects on observing. Whenever a planetary nebula's central star is relatively bright, if you concentrate on the star, the nebula seems to disappear. I have observed this effect also on the galaxy NGC 4685 (L210 in my list). If you concentrate on the unusually bright core of the galaxy, the rest of the galaxy disappears. The reason is that when you concentrate on the central star or galactic core, you are no longer using the eye's outlying rods, which are more sensitive to dim light. The eye's *fovea centralis*, which consists mostly of cones, goes to work on the central star, and so the surrounding nebulosity dims and disappears.

Levy 153   NGC 7293
Helix Nebula or Helical Nebula
Bright nebula
First seen: September 11, 1982
Position (2000.0): α 22 29.6   δ −20 48
Magnitude: 6.0
Distance: 522 light-years
Best seen: in fall; dark sky

Unless you have a dark sky, the Helix will be a very difficult object to find. Perhaps the largest planetary nebula in the sky, it surrounds a central hole with almost a quarter degree of pale, ghostly light. In a dark sky with a wide-field telescope, the Helix is a glorious ghost.

Levy 154   NGC 7009
Saturn Nebula
First seen: November 4, 2002
Position (2000.0): α 21 04.2   δ −11 22
Magnitude: 8.0
Distance: 1,400 light-years
Best seen: in fall; observable in city sky

The Saturn Nebula is at the opposite extreme from the Helix. The Helix is huge, the Saturn tiny. The Helix can be missed entirely because it is so diffuse; the Saturn can be missed because it is not much bigger than a slightly defocused star. But it might be one of nature's finest examples of her sense of humor; plastered in the sky in Aquarius is an object that looks like a ghostly version of Saturn, complete with a ring!

Levy 168   NGC 3242
Ghost of Jupiter Planetary Nebula
First seen: November 4, 2002
Position (2000.0): α 10 24.8   δ −18 38
Magnitude: 7.5
Distance: 1,400 light-years
Best seen: in spring; observable in city sky

> What may this mean,
> That thou, dead corpse, again in complete steel
> Revisits thus the glimpses of the moon
> Making night hideous . . .
>
> —(Hamlet to ghost), Shakespeare,
> *Hamlet* 1.3.51–54, circa 1600

In the depths of night, the Ghost of Jupiter appears in the spring sky. It might look like the planet for which it is named, but it is no planet. It is another gorgeous planetary nebula, looking vaguely like Jupiter as seen through a very poor tele-scope. When Herschel discovered it in 1785, he noted its resemblance to the planet.

Levy 244   NGC 6445
Planetary nebula
First seen: February 8, 2003
Position (2000.0): α 17 49.2   δ –20 01
Magnitude: about 11
Distance: ?
Best seen: in summer; observable in dark sky

This planetary nebula is in the same telescopic field as NGC 6440, a globular cluster; together they make the field unique. The complex is just a short distance north of the diffuse nebula M8. It is large, covering almost an arcminute of sky. I particu-larly enjoy these celestial coincidences of seeing two different types of objects in the same field. When different examples of nature appear together, we get double the bang for our buck.

L245   NGC 6781
Planetary nebula in Aquarius
First seen: March 1, 2003
Position (2000.0): α 19 18.4   δ +06 33
Magnitude: 11.0
Distance: about 2,500 light-years.
Best seen: in summer; observable in dark sky

"Diffuse and interesting" were the words I used to describe this nebula, an almost perfect circle of hazy light that I first saw during the predawn hours of March 1, 2003. I called it inter-esting because it appears enmeshed in a rich field of stars, at least three of them within the nebula itself. (These stars are not really associated with it; they just appear that way in our line of sight.) What looked like a small patch of light through my telescope is actually the expelled outer envelope of an old star; the envelope covers a full light-year in diameter. In photographs through large telescopes, this nebula has a ring shape like the famous nebula in Lyra. The ring feature is lit by a dense, slow wind from the central star, but the fainter parts of the nebula are lit by a faster, less-dense wind.

Levy 286   NGC 1360
Planetary nebula in Fornax
First seen: July 14, 2004
Position (2000.0): α 03 33.6   δ −36 08
Magnitude: about 9; magnitude of bright central star: 11.4
Distance: 978 light-years
Best seen: in fall from the southern United States; much more widely
   from Southern Hemisphere

This radiant planetary nebula has the brightest central star, at least relative to the brightness of the nebula, that I have ever seen. I saw it first on the cold wintry night of July 14, 2004— yes, July was a cold winter night, at an altitude of 12,450 feet in the midst of the Andes mountains in Bolivia. I was trying to search for comets in the constellation of Fornax (the furnace; appropriate for the temperature of the night), a region filled with galaxies of all sizes and descriptions. Suddenly I came across what appeared to be just another galaxy, except this one had an extraordinarily bright star in the middle. It was time to figure out what I was looking at.

As I went from telescope to star and back to telescope and back to star atlas, I became aware of how poorly my brain was functioning at this altitude. I would carefully memorize what was in the field of view—a fuzzy object surrounded by a field of stars—but by the time I got to the atlas a few seconds later, I had completely forgotten what I'd seen in the telescope. So then I memorized the appropriate area of the atlas and went back to the telescope to find that I'd forgotten what was in the atlas. I ended up spending more than an hour identifying it, first with one atlas, then with another atlas with a different scale, and finally with a third atlas. I finally realized that the object at hand was in fact not a galaxy but the planetary nebula NGC 1360. With the altitude and the cold, my brain was unable to function to detect new patterns. Actually, under the best of conditions, I find it frustratingly difficult to compare star patterns on a sketch or in an atlas to what I see in a telescope. With different scales, and especially different orientations (the sketch someone does with a Schmidt-Cassegrain will be "inside-out" compared with what I see in a Newtonian), I can't figure out what's what half the time.

NGC 1360 is huge for a planetary nebula; sizing in at 6.5 arcminutes (a tenth of a degree) of sky, this nebula is five times bigger than the ring. The gaseous shell is oval in shape, which also contributed to my initial thought that I was looking at a weird galaxy. Overall, the time I took to learn that what I was spying was a planetary nebula, not a galaxy, was well worth it.

Levy 33   NGC 1952 M1
Crab Nebula in Taurus: the remnant of the supernova of 1054
First seen: September 1, 1963
Position (2000.0): α 05 34.5   δ +22 01
Magnitude: 8.0
Distance: about 6,500 light-years
Best seen: in fall and winter; observable in suburban or dark sky

I list the Crab at the end of this chapter because even though it signifies the end of a star's normal life, its cause is vastly different from that of all the other planetary nebulae we've discussed. They are the results of gaseous outer envelopes expelled from stars, but the Crab Nebula is the result of a star's utter

Levy 33, M1, The Crab Nebula.
Photograph by Tim Hunter.

destruction. We have records of the time when its supernova's light reached Earth in 1054. In addition to written Chinese records, the supernova might have been recorded by a small group of Anasazi waking in their home under a ledge of rock in what is now northern New Mexico. These forebears of today's Hopi certainly knew the stars. As their eastern sky was showing a sign of dawn, they saw a thin crescent Moon and a strange new star brighter than Venus. One of them, an artist, sculpted that memorable sight of the Moon and blazing star in rock, a petroglyph that survives there to this day.

Nine hundred and nine years later, the remains of that explosion—the Crab Nebula—came my way. Then a patient at Denver's Jewish National Home for Asthmatic Children, I made my ninth attempt to find the Crab. I set up Syncom, the 5-inch f/10 reflector I had at the time, on an athletic field, just north of the asthma home complex. Each one of those nine attempts involved lugging the 5-inch optical tube assembly a quarter-mile distance to where its mount, made of heavy pipe, was waiting.

When I finally achieved success, all I understood was that I had rediscovered the first object in Charles Messier's catalog. I had no idea how that first observation would lead to a lifelong friendship with one of the most important objects in the sky.

THE GALACTIC CENTER
DISTANCE: 26,100 LIGHT-YEARS AWAY

A host, of golden daffodils;
Beside the lake, beneath the trees,
Fluttering and dancing in the breeze.
Continuous as the stars that shine
And twinkle on the milky way,
They stretched in never-ending line
Along the margin of a bay.

     —William Wordsworth,
      "Daffodils," 1804

As our journey continues past the nebulae, the concentration of stars increases dramatically. We depart the spiral arm that contains our

home and head into the vast maelstrom that is the central bulge of stars surrounding the core of our galaxy. What an ethereal place this must be! At the very center there appears to be a colossal black hole whose gravity is so strong that we can actually watch nearby stars swing around it over a timescale of a few years.

In this chapter, we will study just two areas of the sky that are close to our galaxy's center. Both are star clouds—not single objects but vast clouds of stars that mark for us the center of the galaxy in which we live.

Levy 118   The Great Sagittarius Star Cloud
First seen: August 12, 1962
Position (2000.0): α 18 03.4 δ −27 54
NGC 6451: α 17 50.7 δ −30 13
Galactic Center: α 17 45.6 δ −28 56
Magnitude: big and bright
Distance: 10,000 to 26,000 light-years
Best seen: in summer and fall
Stunning crowds of stars and dust lanes; Dark Nebula Barnard 86 is at the west edge of open cluster NGC 6520

On so many nights of comet hunting, my telescope tripped over what I like to call the Great Sagittarius Star Cloud, or abbreviate it as GSSC. And even before I started comet hunting, I remember spending many happy hours mapping it. But on a clear night in August 2002 it stopped me cold. As I moved the telescope across the myriad fields of stars, some laced with strings of dark nebulosity, I realized that I had never really appreciated this sight before. I had seen it, but never before relished its true grandeur.

PROJECT MILKY WAY

How did I begin my studies of the Milky Way? At the Home for Asthmatic Children, my teenage observing sessions included five successful mapping sessions of the Milky Way. The first attempt, on November 10, 1962, failed because that was the night I learned what effect the bright Moon would have on the faint stars in the Milky Way; just a day shy of full, and close to the winter Milky Way in Taurus, its light made it impossible to see all but the brighter stars. I couldn't even make out the Milky Way's outline. On November 24 I tried again, capturing the Milky Way in a much darker sky, and continued on December 2. On March 23, 1963, I charted the rest of the winter Milky Way. Finally, on August 18, I charted the rich central regions.

The purpose of these missions was to understand the extent of the spread of the Milky Way around the sky. What the project did was alert me to the vast differences in the structure of the Milky Way as it spans the sky. I spent much more time getting through the summer Milky Way, from Cygnus, through Aquila and Ophiuchus, down to Scorpius and Sagittarius, than I did navigating the much thinner winter Milky Way in Cassiopeia and Perseus, through Gemini, and all the way south to Canis Minor and Monoceros. The summer Milky Way also seemed divided into two parts. (I'd later learn that the division was a big dark nebula called the Great Rift that made the thinner Milky Way portion look like an exit ramp from a highway.)

As part of my project, I wanted to know why the Milky Way appears to draw a circle around the sky. In fact, from Tucson, just as Cygnus rises in the east, it is possible to trace the whole Milky Way around the horizon. In the answer lies the essence of our position in the galaxy. The circular Milky Way tells us three things: (1) our galaxy is flat, like a CD except with a bulge at the center; (2) its center is in the Sagittarius-Scorpius region of the sky; and (3) (with research stretching out over more than fifty years) we know where we are in the galaxy.

How do we tell that the galaxy is shaped like a disk, or a phonograph record? The Milky Way means two things: the band of milky light and the sum total of all the stars in our galaxy. Our solar system, it turns out, lies in the outskirts of the Perseus-Orion arm of our galaxy. When we look at stars that are away from the Milky Way, we're looking at stars relatively close to us. When we look along the Milky Way, we're seeing the more distant stars in our region of the galaxy. If our galaxy were a giant sphere, the Milky Way would not appear as a specific band of light that crosses the sky.

The second item is easy to see. The Milky Way appears to widen in Scorpius and Sagittarius. But the farther south your observing site is, the more magnificent the central Milky Way appears. From observing sites in the Southern Hemisphere, where Scorpius and Sagittarius can be overhead, the Milky Way is spectacular—bright enough, in a dark sky, to cast a shadow. From such sites it is easy to see that for the Milky Way, here's where the action is. And in the center of this center lies the Great Sagittarius Star Cloud.

Where do we live in our galactic swarm of suns? Our Sun with all its planets—all Earth's history, wars fought and won, political leaders elected to office—all of that takes place at the edge of what's called the "Orion Spur"—an armlike feature that merges with the Perseus spiral arm in the constellation of Cygnus. The Orion feature spurs inward from the Perseus arm, which stretches some 3,000 light-years away. There might be yet another arm farther out, beyond the Perseus arm. As we move closer to the center, we pass through the Sagittarius-Scutum arm, and closer still, the Centaurus-Carina arm. Between that arm and the central hub there might be a bar of stars reaching out from both sides of the center.

OBSERVING THE GSSC

Each spring, several hundred avid observers, hungry for a look at the darkest sky they can find, gather at a ranch in the Davis Mountains near Fort Davis, Texas, for a week of observing. The event is called the Texas Star Party. While there I have heard people exclaim loudly through the night about the beauty of some barely visible distant cluster of galaxies, but, the Great Sagittarius Cloud, all alone, shines majestically over the field, quite ignored by almost everyone.

The GSSC even foiled Messier, which is a bit surprising because he did include its smaller, northern neighbor as No. 24 in his catalog. I think the reason he did not include the GSSC is that his telescope's field of view was too narrow for him to see it all at once. He saw the trees, but missed the forest. So do most modern observers. The GSSC is, after all, not an object but a journey; not a deep sky object to be added to an observing list but an experience to be treasured. The GSSC is the closest thing that the deep sky has to wandering across the Moon with a telescope, crawling into craters, climbing over mountain ranges, and sliding down the Straight Wall.

I could not believe that the night my searching telescope came across the GSSC would launch me on a pilgrimage through star-studded fields interlaced with meandering rivers of dark Barnard nebulae. The easiest way to find the cloud is to point your telescope toward the brightest part of the Milky Way in Sagittarius. However, if you have an automated "go to" telescope, ask it to take you to NGC 6520, a beautiful little open cluster just north of the cloud's center. Along the western edge lies Barnard 80, the most obvious (though not the largest) dark nebula in the cloud. Whichever way you use to locate the cloud, once you're there, stay a while and explore the richness of stars piled on stars. You would think that the cloud would fade out at its edges, but that is not so; there appear to be definite boundaries around it, as well as the dark nebulae that snake around within it.

FINDING THE GALACTIC CENTER

> Towards thee I roll, thou all-destroying but unconquering whale; to the last I grapple with thee; from hell's heart I stab at thee; for hate's sake I spit my last breath at thee.
> —Herman Melville, *Moby Dick*, 1851

Having fallen for the siren call of this region of sky, I thought that the galactic center would probably be in the center of the cloud. Like Captain Ahab sinking the Pequod in search of Moby Dick, I wanted to sail the 26,000 or so light-years with my telescope to the colossal black hole that probably marks the galactic center. As I approached my target I found that the star cloud was not getting richer, but poorer. The galactic center, it turns out, is in the GSSC but not at its center; it lies near the southwest edge near the Scorpius-Sagittarius border. The center is hidden beyond thick dark nebulae, and a powerful telescope is needed to see it. Like Ahab, I would not find our galaxy's Moby-Dickean black hole. Still, how fortunate we are that the entire hub of our galaxy is not similarly obscured and that we have the GSSC to enjoy.

We cannot see the actual center with a small telescope; it is thoroughly obscured by interstellar clouds. However, we can enjoy NGC 6451, an open cluster not related to the center but less than two degrees southeast from it. This cluster is a wonderful object to show at star parties, and to point out that it is the closest deep sky wonder to the center of our galaxy.

THE ACTION AT THE CENTER

In the middle 1950s, a strong radio source called Sagittarius A was proposed as the site of the center of our galaxy. However, the source is mostly invisible optically. More recent observations have moved the center slightly to another very dense radio source called Sagittarius A* (A prime).

Some 26,100 light-years away, Sagittarius A* has received intensive study in recent years. The radio source and its surrounding stars cannot be seen in visible light because of obscuring dust. In infrared light, the galactic center reveals itself as a very complex place where stars race around the radio source quickly. The closer the nearby stars are to the center, the faster they orbit it; one star, only a light-week away from the source, exhibited visible motion in only two years of observation.

From these orbital characteristics, Andreas Eckart and his colleagues have suggested that Sagittarius A* is a "supermassive" black hole that is some 2.6 *million* times the mass of the Sun.[1] It seems that the peaceful, docile structure of a galaxy that we believed years ago we were living in was wrong. The Milky Way galaxy is a pretty exciting, energetic place!

Levy 119   NGC 6603 M24
Small Sagittarius Star Cloud
First seen: August 13, 1963
Position (2000.0): α 18 16.9   δ −18 29
Magnitude: about 3
Distance: 9,400 light-years
Best seen: in summer and fall

M24 is not really a star cluster, as Steve O'Meara notes in his excellent deep sky companions book *The Messier Objects*. "Commonly called the Small Sagittarius Star Cloud, M24 is a virtual carpet of stellar jewels, laid out across 330 light-years of space." Through his telescope, O'Meara reports that "no sight in the visible universe shares M24's mystical qualities."[2] No sight indeed, with the possible exception of the LSSC. Within M24 there is a real star cluster, NGC 6603. It is fat and dense with many faint stars and a stream of stars that ends with faint planetary nebula NGC 6567. Also, Barnard 92 is a Bok Globule nearby.

HOW WE DISCOVERED
THAT OUR GALAXY IS SPIRAL

Most of the nebulae that we see in the night sky exist in our galaxy's spiral arms and not in its central hub. As our galaxy rotates about its center, it sends out "density waves" that propagate throughout the spiral arms. These ripples have the effect of gathering the galactic supply of gas and dust and compressing them into areas that spiral out. The results of these waves, then, can help us define the galaxy's spiral shape.

Not yet understanding the meaning of density waves, in the early years of the last century, Kapteyn envisioned a small galactic system with the Sun in the center. Around 1920 Harlow Shapley completely rewrote this idea. His studies of the

Levy 119, M24, a complex of dark lanes and bright stars.

globular clusters, which he saw dotting the galaxy's outskirts, proved that the galactic center was far from the Sun and that our galaxy is huge.[3]

The idea that our galaxy is spiral came up during the 1930s, and one of its most dramatic proponents was Bart Bok. Many astronomers strongly suspected the Milky Way of being a spiral galaxy, even though the patterns of the arms were far from proven at that time. While enjoying a cruise through the Panama Canal, Bok had a magnificent view of both the Milky Way's northern and southern stretches, and he could visualize the sweep of a possible arm. As he looked up at the Milky Way, he envisaged a neat spiral structure to our galaxy, a spiral arm beginning in Cygnus, which moved south through the Sun into the southern Milky Way constellation of Carina.

This might have been a fanciful prognostication, but the idea of how the stars were distributed in space was not. Bok imagined a Milky Way that spiraled out like a monstrous pin-wheel. That night the presence of this arm made sense as he stared at the Milky Way, but the idea did not survive the rigor of astronomical observation. The Carina-to-Cygnus idea lasted long enough to appear as the closing words of his seminal work *The Distribution of the Stars in Space*, in which he explored our galaxy from the point of view of how the stars in our galaxy were arranged. "The observer in the tropics," he wrote based on his own observing experience on the cruise, "should not find it difficult to accept as a working model for our Milky Way system one with a distant center in Sagittarius and in which a spiral arm passes from Carina through the sun toward Cygnus."[4] With the benefit of hindsight, we now know differently. There is probably an arm in Carina (known as the Sagittarius-Carina arm), and there is an arm in Cygnus, but they do not connect through the Sun. Instead, the solar system belongs to the Orion-Perseus arm.

During World War II, the race to understand the size and

shape of our galaxy gathered momentum. A big reason: Walter Baade, a German astronomer at Mount Wilson, misplaced his papers after immigrating to the United States. When the United States declared war against Germany in 1941, Baade was classified as an enemy alien and held under virtual house arrest. However, by listing his primary residence as Mount Wilson Observatory, he was permitted to remain there throughout the war. With most of the astronomers doing war-related research, and nighttime blackout conditions imposed on nearby Los Angeles, Baade had several years of uninterrupted observing time on the 100-inch Hooker telescope, then the world's largest.

During these years, Baade studied the spiral patterns of nearby galaxies. He defined two "populations" of stars, a specific type of population that makes up the spiral arms and a different type that inhabits the central bulge. He also identified regions of hydrogen gas that accompany the stars in the arms. After the end of the war, Baade understood the significance of what he had observed, and he corresponded with Bart Bok on how his studies of the Andromeda Galaxy could relate to the Milky Way. "From my studies of the Andromeda nebula," Baade wrote Bok on February 8, 1949 (and notice his use of the term nebula instead of galaxy!), "I would bet that the absorption for the 4 Cygnus Cepheids is due to the fact the line of sight (from them to us) runs in the absorption-free—or at least absorption-poor—space between two neighboring spiral arms."[5] What he meant was that the light from these four Cepheid variable stars in the constellation of Cygnus was passing through the relatively dust-free areas between two of our galaxy's spiral arms as it travels to our telescopes.

Our scene now shifts ahead two years. By 1951 astronomers were ready to make the intellectual leap to understand that the galaxy in which we live has spiral arms and to back that up with observational evidence. William Morgan, Stewart Sharp-

THE GALACTIC CENTER 147

less, Donald Osterbrock, and all of Yerkes Observatory followed up on Bok's ideas about the distribution of stars, which had been expanded upon by Baade. In studying the distribution of stars of different populations in the solar neighborhood, they detected evidence of two spiral arms, which they called the Orion and Perseus arms, plus part of a third called the Sagittarius arm.

enter the big dishes

The advent of radio astronomy, with the dishes of radio telescopes peering into the sky, offered a totally new way to confirm and expand the discovery of a spiral structure. While optical telescopes use mirrors or lenses to see deeply into the sky, radio telescopes employ large dish-shaped antennae to hear it. The dishes have to be big, yet all the radio energy they have ever collected, Carl Sagan told us, is less than the energy of a single snowflake. In 1951, using a small pyramid-shaped horn antenna mounted on one roof of Harvard's civics building,[6] Harvard physicists Harold I. Ewen and Edward M. Purcell detected radiation from neutral hydrogen atoms at the 21-cm wavelength as a radio signal from the Milky Way.[7] Before the advent of radio telescopes, the galaxy's shape lay hidden behind a dark veil of interstellar dust that optical telescopes could not penetrate. But radio telescopes "see" a different wavelength of sky, and through them, the Milky Way's spiral shape could be mapped. The spiral arms are traceable by observing where hydrogen is especially concentrated. Not only could the Orion and Perseus arms be confirmed, but the arms could also be extended much farther out, beyond the dark nebulae that blocks the view of the optical telescopes. Bok's old mentor Jan Oort, and also van de Hulst, then went on to publish a 21-cm wavelength map of the galaxy, showing its detailed spiral structure.

By the 1970s and early 1980s, Bart Bok loved to explain how "we used to think that Andromeda Galaxy [the nearest major galaxy] was twice as large as the Milky Way. Now we think that Andromeda still is bigger," he emphasized, *"but not by as much."*[8] Our Milky Way seemed more complex as well. For example, it was previously thought that the galaxy rotated "nicely and politely" like the Earth does. But now, he would go on, astronomers recognize that different parts of the galaxy rotate at different rates, the inner section much more rapidly than the outer.

With his decades of observing experience with the Milky Way, Bart Bok developed a marvelous appreciation for how things are ordered in the sky. "When you are in an observatory at three o'clock in the morning," he told his students, "stop your photograph. Stop your photometer. Walk away from the telescope. Walk down the stairs. Walk out the front door. Now walk twenty paces—no more, no less. Then stop—and look up at the sky—just to make sure you are making bloody sense."[9] There has been no better advice. Never forget to step back from your telescope and just look up to enjoy the full panorama of our home galaxy.

CHAPTER 9

GLOBULAR CLUSTERS
DISTANCES: TENS OF THOUSANDS TO MORE THAN A HUNDRED THOUSAND LIGHT-YEARS AWAY

She is the most beautiful creature in the universe; and yet she is mistress of such noble, elevated qualities, that though she is never from my thoughts, I scarce ever think of her beauty, but when I see it.
—Henry Fielding, *Tom Jones*, 1749

I can see 47 Tucanae only when observing from the Southern Hemisphere, so between visits, I forget just how gorgeous this globular cluster truly is. It is a stupendous example of one of the richest, most graceful structures the night sky has to offer. Moving farther into space, we now turn away from the Milky Way's center and spiral arms, and out to its halo. There we find some the oldest and largest structures in the Universe: globular clusters. They are exquisite. Each one is home to tens of thousands of stars. These clusters are among the most stunning objects in the galaxy, especially the Southern Hemisphere's sparkling treasures of Omega Centauri and 47 Tucanae. They also have been invaluable in helping us to understand the size and age of the Universe.

Levy 3 NGC 6341 M92
Globular cluster in Hercules
First seen: July 4, 1966
Position (2000.0): α 17 17.1   δ +43 08
Magnitude: 6.4
Distance: 26,000 light-years
Best seen: in summer; observable in city sky
Nice globular cluster; thick, well-defined nucleus
Shapley class IV

A few miles away from the Adirondack Science Camp, where I met M92, stands a beautiful oval-shaped mountain with a solid face of exposed metamorphic rock. Called Pocomoonshine, the mountain is a famous rock climbing site, but it is also a great mountain just to admire; under the light of a bright Moon, the rock face glows with an unforgettable pale light. On Independence Day evening 1966, a bright gibbous Moon, only three days past full, created just such a glow both on Pocomoonshine and in the sky above the camp. It was in that moonlight that I was introduced to M92 and the chain of stars that seems to accompany it, but which is really closer than it looks.

Levy 3, M92.

WHAT IS A GLOBULAR CLUSTER?

Even after Abraham Ihle discovered the fuzzy patch of light in the constellation of Sagittarius in 1665 that we now call Messier 22, globular clusters were not an especially interesting part of our astronomical studies. Twelve years later in St. Helena, Edmond Halley found a large oval-shaped globular cluster now called Omega Centauri, and in 1714 he recorded the great cluster in Hercules. At the end of the eighteenth century, William Herschel began his survey of the northern sky, his goal being to list everything his telescopes could find. He described some objects—the ones that turned out to be globular clusters—as shining with "a mottled kind of light" which later and better telescopes resolved into the many stars that make up the cluster.

Globular clusters are masses of stars that typically lie on the outskirts of our galaxy; they are among the best objects to view from a city sky. We know of 150 globular clusters in our galaxy, and there are probably another hundred whose light is blocked by the great intervening clouds of dust in space. While some clusters are now near the center of the galaxy, others inhabit the lonely part of space, far away from the richness of stars in our galaxy. But a cluster passes through the dense regions of the galactic plane twice each orbit. The dense layers of dust disrupt the galactic plane, which the cluster passes through. The pressure of the dust can be strong enough to push some stars out of their clusters entirely. The area around our galaxy, called the galactic halo, is littered with old, lonely stars that, eons ago, enjoyed membership in outlying globular clusters.

Because the globular clusters are concentrated in the half of the sky closest to the center of the galaxy, the summer sky, with its brilliant Milky Way, contains most of them. By contrast, the winter sky offers the diminutive Messier 79, south of Orion's Belt.

The stars that form globular clusters are quite different from ordinary stars. The brightest are yellow or red giants. The clusters are also immensely old—some calculations show that Messier 13 has been around for some 13 billion years. The Universe itself is estimated to be 13.7 billion years old. If that figure is correct, globular clusters are among the oldest structures in the Universe. Globular-cluster stars tend to be devoid of heavy elements, and they are not surrounded by gas and dust. Hence the stars in the clusters must have been formed before the gas surrounding them was enriched by heavy elements from supernovae.

But these clusters may be also forming new stars. Some globular clusters have unusual stars called blue stragglers. The mystery of these stars is that they contradict the established theory that all the stars in a globular cluster formed at the same time, very long ago. These young stars spin quickly—the Hubble Space Telescope recently provided evidence that one is spinning once in less than a day, some seventy-five times faster than the Sun. The stragglers may be formed from mass exchanges from nearby stars in the crowded core of a globular cluster.

Levy 4   NGC 6254 M10
Globular cluster in Ophiuchus
First seen: September 13, 1964
Position (2000.0): α 16 57.1   δ –04 06
Magnitude: 6.6
Distance: 20,000 light-years
Best seen: in summer and fall; observable in city sky
Shapley class VII

Messier 10, at 20,000 light-years distance, is one of the closer globulars to us. Its stars appear somewhat typically concentrated toward its center. (Although globular clusters tend to have a similar appearance based on the enormous gravitational pull of their stars, they vary in appearance depending upon how closely their stars are concentrated.) Shapley classified them on a I to XII scale, where a type I cluster is densely concentrated, and a XII has virtually no concentration at all, its stars being loosely distributed. The complete Shapley scale is as follows:

     I: high concentration toward center
    II: dense central concentration
   III: strong inner core of stars
   IV: intermediate rich concentration
    V: intermediate concentration
   VI: intermediate [less concentration than V]

  VII: intermediate [even less concentration than V]
 VIII: rather loosely concentrated toward center
   IX: loose toward center
    X: loose
   XI: very loose toward center
  XII: almost no concentration in center

Levy 6   NGC 6229
Globular cluster in Hercules
First seen: Bastille Day 1966
Position (2000.0): α 16 47.0   δ +47 32
Magnitude: 9.4
Distance: 100,000 light-years
Best seen: in summer; observable in city sky
Shapley class IV

Bastille Day, celebrated as the day the old Paris prison was stormed and which triggered the French Revolution, is not normally a day that would have much significance for astronomy. But in Charles Messier's life, it did. With the start of the French Revolution, Messier lost the pension that allowed him to live while he spent his time hunting for comets. Virtually penniless, he even had to borrow oil for his observing lamp from Lalande, one of his friends.

This cluster, often forgotten since it is the third-brightest cluster in Hercules (after M13 and M92), is an unusually lovely sight, thanks to its location near two foreground stars, with which it forms a near perfect equilateral triangle. Thus I usually don't forget our encounters, of which there have been many since that Bastille Day. The most memorable of those meetings took place on the night of November 13, 1984 (see chapter 5); about fifteen minutes later, I found my first comet.

Levy 6, NGC 6229.

Levy 12   NGC 7078 M15
Globular Cluster
First seen: August 23, 1966
Position (2000.0): α 21 30.0   δ +12 10
Magnitude: 6.4
Distance: 34,000 light-years.
Best seen: in fall; observable in city sky
A favorite globular
Shapley class IV

> In the mixture of starlight and cloud-reflected sunlight in which the desert world is now illuminated, each single object stands forth in preternatural though transient brilliance, a final assertion of brilliance before the coming of night . . .
> —Edward Abbey, *Desert Solitaire*, 1968

Levy 12, M15.

In the magic that joins earth and sky, it is somehow satisfying to look at the dying light and see the dim outline of a mountain top and then look into the night for a view of the mighty globular cluster M15. I have admired this particular globular on many nights, with a special fond memory of October 14, 1964, a red-letter day in my fledgling astronomical career. An event called Star Night, held at Westmount Park, would be my first opportunity to show an object in the sky to the public. Pegasus and I would be part of a team of several dozen observers with their telescopes. I was even assigned an object to observe: Messier 15, a beautiful globular cluster in the 8-inch telescope's namesake constellation of Pegasus.

This was not my first Star Night. In fall 1960 I walked nervously to an earlier Star Night; with thousands of people in the park, I wondered if I'd even get to look through a telescope. I did, and I also got to ask a question! That evening each participant wrote a question on a card and handed it to the registration desk. I did, and about twenty minutes later the moderator, physicist T. F. Morris of McGill University, asked my question to the crowd: "Do all stars belong to galaxies?"

"The Universe is organized gravitationally into galaxies," Dr. Morris explained. Some stars may belong to the globular clusters that orbit galaxies, but it would be most unlikely to have an intergalactic wandering star all alone in space. Messier 15, he might have added, is a fine example of a globular cluster visible in the autumn sky.

By the following September, at thirteen years old, I had become a part of the group that sponsored Star Night, the Montreal Centre of the Royal Astronomical Society of Canada. With great excitement I asked if the center could use the help that I might provide with my new 3.5-inch Skyscope, Echo. "No," I was told. "We can't."

Still enthusiastic about helping out with my little scope, I persisted. "Well, if there is anything I can do . . ."

"Listen David," I was told in no uncertain terms, "we have a rule here about people not being allowed to join the center until they are sixteen years old, and it's a good rule. When I say we can do without your help, I mean it."

In my mind I decided then and there to drop out of astronomy forever. Fortunately that decision didn't hold; by the time I reached my sixteenth birthday in 1964, I was still enthusiastic, still wanting to volunteer to help out at Star Night. This time the group was far more receptive and assigned me the mighty globular cluster M15. As the magic evening approached, I spent every clear night getting familiar with where it was in the sky.

After school that afternoon, I met Carl Jorgensen and Bryan Rawlings at the CFCF television studio, where I was to make an appearance on a show called *Magic Tom*. As the host approached us seconds before airtime, dressed as a clown, he commented to us, "The things you gotta do for a living!" He was supposed to discuss Star Night with the youngest of the telescope operators. As hundreds of people gathered at the park a few hours later, so did the clouds. As we waited, there was a Q&A session, and people asked me about my telescope and what I would show them if only it would clear. I had the chance to explain how Messier 15 was some 30,000 light-years away and how it had been cataloged by Messier as an object he had come across during his search for comets. At 9:30 PM the sky cleared, and I got to show off my telescope and a distant globular cluster in space. I felt as proud as if I were showing off my own child.

M15 encountered me again while I was comet hunting on August 23, 1966. This time there were no crowds, just my old friend M15 and me. At this pristine location deep in the Adirondack Mountains, the appearance of M15 was vastly different; instead of a small round mottled nebula, I saw the cluster in its full glory, a cluster of stars projected against the inky backdrop of space.

In recent months I've had a new type of reacquaintance with my old friend. I go outdoors, spend a few minutes getting dark adapted, and then look toward the cluster, without telescope or binoculars. I've actually seen it with the naked eye! May this beautiful cluster, which has been known since Jean-Dominique Maraldi discovered it in 1746, inspire other young people as it inspired me.

Levy 21   NGC 6838 M71
Globular cluster in Sagittarius
First seen: September 2, 1964
Position (2000.0): α 17 17.1 δ +43 08
Magnitude: 8.3

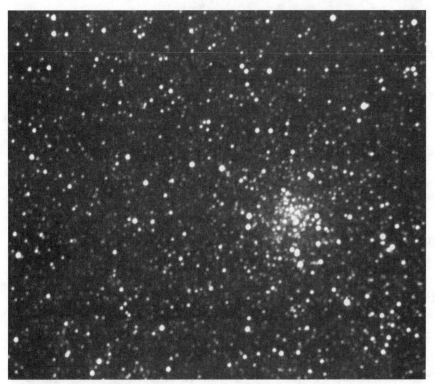

Levy 21, M71, one of the "loosest" globulars in the sky.

Distance: 13,000 light-years
Best seen: in fall; observable in city sky
Shapley class XII+

Although this object counts as a globular, it is really an example of an open cluster at the edge of a globular. Philippe de Chéseaux is likely its discoverer. He saw it in 1746, two years after his great comet. Its stars are well spaced apart. It is easy to resolve because, at a distance of 13,000 light-years, this small cluster is relatively close to us.

Levy 23   NGC 5024 M53
Globular cluster

First seen: May 22, 1964
Position (2000.0): α 13 12.9   δ +18 10
Magnitude: 7.7
Distance: 56,000 light-years
Best seen: in spring; observable in city sky
Shapley class V

This is a most remarkable globular cluster. However, because of its location within the Coma-Virgo cluster of galaxies, we tend to ignore it in favor of the rich field of galaxies. Just one degree to the southeast is NGC 5053. The two globulars are a remarkable celestial coincidence; brighter M53 is 7,000 light-years farther away.

What would it be like to live in this globular cluster? With so many stars, our twilight sky might never darken completely. And then off in one direction would be the soft light of NGC 5053, shining like a bright ball of light in its sky. It would be extraordinarily beautiful. As for the people living on a world circling a Sun in M53, they would have a much more dramatic view of NGC 5053 than we do.

Levy 25   NGC 1904 M79
Globular cluster in Lepus
First seen: January 1, 1965
Position (2000.0): α 05 24.5   δ −24 33
Magnitude: 8.0
Distance: 42,000 light-years
Best seen: in winter, observable in city sky; rare winter globular
Shapley class V

To everything there is a season, Ecclesiastes says, and that's true for the sky as well. Each season offers its special features. Summer is the time for the Milky Way, spring for the galaxies. But until Bart Bok pointed it out to me, I never realized that it was also true for globular clusters. Globular clusters are the most distant objects generally visible from light-polluted cities, and it is remarkable that so few of us realize how disproportionately spaced they are in the sky. "An astronomy trivia question for you," Bok began one day as we were chatting together back in 1979. "Is there a globular cluster that is *best* seen on winter nights in the Northern Hemisphere?" Bart didn't let me think about his question for long. "It's M79," he winked, "just below O'Ryan, the Irish Constellation."

New Year's Day 1965: we always hope that a new year will bring us good and exciting times. This particular year started off very well, with a look at the rising numbers of sunspots after the sunspot cycle passed its minimum in 1964. On New Year's night, I had my first acquaintance with Messier 79. Discovered by comet hunter Pierre Méchain in 1780, Messier 79 is a must-see for any winter star party. From the bright city sky of Montreal that night, it was a bright, fuzzy-looking object. From darker locations, I later was able to see more of the "starfish" structure of this cluster's outlying regions of stars.

Levy 97   NGC 5139, Omega Centauri
Globular cluster in Centaurus
First seen: May 9, 1980
Position (2000.0): α 13 26.8 δ –47 29
Magnitude: 3.7
Distance: 17,000 light-years
Best seen: in spring from the southernmost United States, but even
    better seen in Southern Hemisphere; observable in city sky
Shapley class VIII

Only two globular clusters in the sky are so bright that they are given official star names: Omega Centauri, with its Bayer letter, is one; and 47 Tucanae (a Flamsteed number) is the other. Which cluster is superior? Because it is more concentrated, 47

Levy 97, Omega Centauri.
Tim Hunter Photo.

Tucanae appears brighter. Omega Centauri is bigger, and at magnitude 3.7 it is the brightest globular cluster in the sky, but just by a tiny amount, over 47 Tucanae. Both clusters are worth a trip to the Southern Hemisphere

April 1-5, 1988   Steve and Donna O'Meara. and that omnivorous comet catcher Levy
4i David, just for the record I did morning and evening
Nova patrols from your L.C. Peltier Observing Station (on your sunroof).
I did not find any, but I had a "close call" in Puppis, east of where
Nova Puppis 1942 lies. As you know, since you were present, I viewed
Omega Centauri rising above the Empire Mtns. (Fagan Mt.). - "The
revelation was sublime." Thus would sayeth G. P. Bond. Donna
did a few comet sweeps with the 6" and stumbled upon numerous clusters in
Puppis. I showed you + Donna the Eskimo nebula in Gemini, M 106 in U. Maj,
some NGC galaxy in the front paw of U. Maj; M 81-82, and other
this and that.

O'Meara Centauri Rising: A Comet hari parent
as seen through Minerva, David's 6-inch

Stephen James O'Meara

Levy 97, "O'Meara Centauri." Sketch by Stephen James O'Meara from our home, April 1988.

to see, but if you live in the southern part of the United States, you can see Omega Centauri low in your southern sky. It is visible to the naked eye on a dark night and clearly visible through binoculars.

Omega Centauri is the largest globular cluster in our galaxy. More than 280 light-years from one side to the other, it houses several million stars. It is *ten times* more massive than most of the other big globular clusters, and, as Steve O'Meara notes, it is almost as big as a small galaxy. Although it has been observed as a starlike object probably for thousands of years, Edmond Halley was the first to record its nonstellar appearance while he was at St. Helena. (It is interesting how so many deep sky discoveries have come from the eyes of people primarily known for comets!) In 1827 James Dunlop of New South Wales first recorded its true nature as a cluster of stars.

Could this cluster be *more* than a globular cluster? If the South Korean astronomer Young-Wook Lee and his team are correct, Omega Centauri has undergone a long, two-billion-year period of star formation. This behavior is more typical of a galaxy nucleus than of a globular cluster, and so it is possible that Omega Centauri is actually all that is left of a galaxy that collided and merged with the Milky Way long ago. The Milky Way may be twice guilty of galactic cannibalism: the Sagittarius Dwarf Galaxy is now being ingested, and globular cluster Messier 54 may be the remains of its nucleus.[1]

Levy 104   NGC 104, 47 Tucanae, NGC 121, and Lindsay 8
Globular cluster in Tucana, near but not part of the SMC; NGC 121 and
    Lindsay 8 are part of the SMC
First seen: June 18, 2001 (47 Tucanae)
Position (2000.0): α 00 24.1   δ −72 05
NGC 121: α 00 26.7   δ −71 32; Lindsay 8: α 00 25.0   δ −72 45
Magnitudes: 3.9; 11.2; 12.5
Distances: 15,000 light-years; 176,000 light-years

Best seen: only in Southern Hemisphere
Shapley class III (47 Tucanae)

> I could be bounded in a nutshell and count myself a king of
> infinite space, were it not that I have bad dreams.
> —William Shakespeare, *Hamlet* 2.2.253–56, circa 1600

Two globular clusters in the same field of sky: one is bright enough to be visible without a telescope; the other shines from another galaxy. Forty-seven Tucanae is simply stunning through a large telescope while NGC 121 is barely visible through one. Forty-seven Tucanae is one of the closest globulars to us; NGC 121 belongs to another galaxy, the Small Magellanic Cloud.

First, 47 Tucanae: You can't just observe this cluster, or study it, through a 16-inch telescope; you bask in it. When I saw it through a telescope that big at the Mamalluca public observatory in Vicuña, Chile, the light bluish glow from the cluster's center seemed so close I felt that I could warm my hands with its light. Observing it with a 9-inch refractor, Steve O'Meara writes that it "stole my breath. The cluster literally burned from within. A tight 2.8'-wide band of topaz light sizzled like an electric flame while thousands of sparks illuminated sphere upon sphere of starlight, which diffused outward from the cluster's center."[2]

Now to NGC 121 and Lindsay 8: O'Meara points out how close the far-off NGC 121 is to 47 Tucanae—just half a degree north-northeast of the center, and just 10 arcminutes from the edge of the bigger cluster's outer halo.[3] One of the faintest globulars, it is only 11.2 magnitude. If it were placed at the same distance as 47 Tucanae, it would appear only a third as large.

Lindsay 8 was first catalogued by Eric Lindsay, who worked closely with Bart Bok and Harlow Shapley at Harvard, and later at the Boyden station in South Africa, to study with Bok. But even though the other clusters are much farther away, a part of another galaxy, they are part of the second-closest galaxy to our own. By looking at these three globulars we get a sense of distance in the universe; by understanding that even with NGC 121 and Lindsay 8 we're in our own neighborhood, we can appreciate that vastness.

Levy 105   NGC 362
Globular cluster in Tucana, on opposite side of Small Magellanic Cloud
    from 47 Tucanae
First seen: June 18, 2001
Position (2000.0): α 01 03.2   δ −70 51
Magnitude: 6.4
Distance: 27,700 light-years
Best seen: only in Southern Hemisphere
Shapley class III

As close as this cluster is to 47 Tucanae, it always seems to be in the "shadow" of its famous neighbor. Actually one of the brightest globulars in the heavens, NGC 362 is on the north side of the Small Magellanic Cloud, a galaxy of which neither it nor 47 Tucanae are a part. Discovered by James Dunlop from New South Wales in the 1820s, the cluster is an easy binocular target. Its Shapley score of III—indicating a dense concentration of stars near its center—is the probable reason why the cluster has survived. Globular clusters orbit the center of our galaxy in various types of orbits; NGC 362 periodically dips to within some 3,000 light-years of the center of the Milky Way. If the cluster were not so well held together, it would never have survived about one hundred passes through the galaxy's densest regions.[4] The Milky Way's halo, a large region around the galaxy, is littered with the starred remains of globular clusters that were not so lucky.

Levy 117   NGC 6093 M80
Globular cluster in Scorpius
First seen: July 14, 1963
Position (2000.0): α 16 17.0   δ −22 59
Magnitude: 7.3
Distance: 33,000 light-years
Best seen: in summer; observable in city sky
Compact globular cluster; looked fuzzy when sighted in 1980s near the
    eastern horizon
Shapley class II

What does M80 have to do with a total eclipse of the Sun?

I went nuts trying to find M80 during one of the busiest periods in my teenage life. From my diary at the asthma home on July 13, 1963, I wrote how, while walking, I went through in my head every stage of the partial and total phases of the eclipse of the Sun, an eclipse I hoped to see one week later and two-thirds of a continent away. A few hours later, however, "I couldn't find M-80 with Syncom [the 5-inch f/10 reflector I had at the time] tonight—I tried and tried and tried, but I couldn't find it."

July 14: "Eclipse plans completed. T-6 days. That is all. I read *Sky & Telescope* today. It was mainly cloudy when I went out to find M-80. But it cleared a bit later, and I [began searching] at 9:01. I found M-80 slightly before 9:02 and kept it [in the field of the eyepiece] until about 9:04. What a fast success!"

Were it not for my observing log, nothing about my successful meeting with M80 would be on record. I "discovered" it for myself that evening, just as Charles Messier had discovered it for himself, and for all humanity, in 1781. But both experiences taught me something about the need for patience, patience, and more patience when we come head-to-head with the cosmos. There was no question of my frustration when I failed to find M80 that first night, but an instant success the next night has become part of the observing game one plays with nature. Or maybe I suddenly realized that M80 really is an easy target, about halfway between Antares and Beta Scorpii— at that time the two brightest stars in the Scorpion. (In recent years Delta has brightened remarkably, and at the time of this writing it, not Beta, is the constellation's second-brightest star.) Nature helped in another way that night; because the cluster is Shapley II, with a very dense concentration toward the center, it was easy to see from a city sky.

Another fact I didn't know in 1963: we were just past the centenary of a very rare event—the discovery of a nova, or exploding star, in a globular cluster. On May 21, 1860, the Dutch astronomer Arthur von Auwers discovered T Scorpii as it flared from M80's core. Outshining the entire cluster in brightness for less than a week, it then faded.

Levy 147   NGC 6638
Globular cluster
First seen: October 24, 2002
Position (2000.0): α 18 30.9   δ –25 30
Magnitude: 9.2
Distance: 26,000 light-years
Best seen: in summer; observable in city sky
Shapley class VI

When my friend Dean Koenig suggested I look at this diminutive cluster, I thought he meant either of its two brighter

cousins, M28 or M22. But this faint globular cluster, which forms one end of a triangle of three interesting globular clusters, teaches us something about

Levy 147, NGC 6638.

the distances to these clusters. NGC 6638 is just a little more than a degree southeast of its brighter neighbor M28, and southwest of the monster globular M22, one of the largest in the sky. However, it is about 7,000 light-years closer to the galactic center than M28, and 16,000 light-years closer to it than M22, and consequently more subject to the gravitational stresses that could tear it apart. Seeing these three clusters together gives us a sense of distance in the Milky Way. (At the end of this chapter, we will peer even closer to the center at two globular clusters even farther from us and closer to the center of our galactic home.)

Levy 195   NGC 2419
Globular cluster
First seen: November 13, 2002
Position (2000.0): α 07 38.1   δ +38 53
Magnitude: 10.3
Distance: 300,000 light-years
Best seen: in spring; needs dark sky
Lynx globular cluster; this is Harlow Shapley's Intergalactic Wanderer
Shapley class II

> This most excellent canopy the air, look you, this brave o'er-hanging firmament, this majestical roof fretted with golden fire . . .
>
> —Shakespeare,
> *Hamlet* 2.2.299–302, circa 1600

NGC 2419 is just north of the bright star Castor in Gemini. It first revealed itself as a globular cluster to the sharp eyes of Carl Lampland at

Levy 195, NGC 2419,
Shapley's "intergalactic wanderer."

Lowell Observatory. This cluster is so far out of our galaxy —more than 300,000 light-years away—that Harlow Shapley thought it to be totally free of its gravitational pull, so he called this cluster an "intergalactic tramp." The cluster, it turns out, is probably connected to our galaxy, looping around its center in a very eccentric orbit.

HENRIETTA LEAVITT, HARLOW SHAPLEY, AND THE DISTANCES TO GLOBULAR CLUSTERS

To tell the story of how globular clusters helped us determine the size of our galaxy, we must go back more than two centuries, to October 1784, when the English teenage deaf-mute John Goodricke (see chapter 4) discovered that Delta Cephei was changing in brightness. By the age of nineteen, Goodricke had discovered the variations of three stars, but Delta Cephei, it later turned out, was special. Its pulses were not caused by the periodic eclipsing of the star by a fainter companion. Instead, this star by itself varied periodically, and precisely, over a period of 5.37 days. The star changes brightness as it expands and contracts on a regular basis; as it expands it fades, and as it contracts it brightens. Cepheid variables, precise as clockwork, have been found to exist all over our galaxy. Harvard's Solon Bailey had noted large numbers of them in the globular clusters. After 1902 it was Leavitt's turn, and despite interruptions caused by illness, she found them in the Magellanic Clouds as she continued her work.

To find these variable stars Leavitt used the method of "superposition." She would place a negative from a photographic plate taken on one date on top of a positive print of a photograph taken on another date. The black and white images should have coincided, canceling each other out. If they did not, Leavitt suspected the star was variable and would then repeat the process with a different set of photographs to confirm her discovery.[5] In 1904 she published her first list of variable stars discovered in the Magellanic Clouds, using this procedure. They included 152 variable stars in the Large Magellanic Cloud (LMC) and 59 in the Small Magellanic Cloud (SMC). Concentrating on the small cloud in the next year, she found more than eight hundred variables there. These discoveries were published by the observatory as they occurred, and Edward Pickering received several letters that year praising Leavitt for her enthusiasm and ability to discover variable after variable. One of these letters even suggested that Miss Leavitt be nominated for the Nobel Prize.

In 1908 Leavitt published an exhaustive list of 1,777 variable stars. From this huge list she focused on sixteen Cepheid-type variables whose periods ranged in time from a short 1.25 days to 127 days. She also reported, somewhat tentatively at the time, that "it is worthy of notice that the brighter variables have the longer periods."[6] Although she had no idea at the time, Henrietta Swan Leavitt was onto one of the fundamental discoveries in astronomical history.

In 1913 a young astronomy student Harlow Shapley set out for the Mount Wilson Observatory in California. Mount Wilson was at the astronomical frontier. It had a mighty 60-inch telescope, and an even mightier 100-inch was being planned. Shapley began observing distant variable stars, especially those in globular clusters. Since all the stars in the Magellanic Cloud are about the same distance from us, Shapley concluded from Leavitt's work that the stars display a correlation between their periods of variation and their average magnitudes. He turned this relationship into an astronomical yardstick for measuring distances.

Shapley observed through the 60-inch Mount Wilson reflector, one of the finest optical telescopes ever built. He discovered Cepheid variable stars in globular clusters and that those stars showed the very same period-luminosity relationship that Leavitt had discovered earlier. By this time he had both confirmed Leavitt's work and expanded it to include closer globular clusters. He saw the enormous potential of Leavitt's discovery: if the period of variation gave a clue to the star's real brightness, then by comparing the absolute with the apparent brightness, one could calculate the distance to the star.

Shapley proposed that if the absolute magnitudes of the Cepheid stars in globular clusters and the Small Magellanic Cloud could be determined, then one could calculate the distance to these places. (A star's absolute magnitude is its brightness regardless of its distance from us; its apparent magnitude is its brightness as we see it, a factor influenced by both the star's absolute magnitude and its distance.) The average apparent magnitudes in one cluster would be brighter or fainter than in another cluster; because of this, Shapley was able to arrange the clusters by their relative distances from us. However, to know how far these clusters were in light-years, Shapley needed to work the period-luminosity relationship backward.

He needed to determine the absolute magnitudes to some Cepheids, even just a few, in our own galaxy, by a different process. Why could he not use the period-luminosity relation within our own galaxy? After all, the Milky Way galaxy contains many Cepheid variable stars. Although these stars are all closer than those in neighboring galaxies, they are at different distances from us. More important, since we are viewing them from within our own galaxy, these stars can appear brighter or fainter because of the thickness of interstellar dust that lies between each variable and Earth. A star in Cepheus, for example, might appear to have the same brightness as a star in Scorpius, but one might be dimmed more by interstellar dust.

The period-luminosity relation is hard to apply within our galaxy because the intervening dust prevents us from knowing the star's true luminosity. By studying the variable stars in the closest of the other galaxies, Leavitt avoided this problem. *All the Cepheids in the Small Magellanic Cloud are about the same distance from us*—the distance of that galaxy. They are all affected equally by whatever interstellar dust lies in our own galaxy on the line of sight from them to us. The Small Cloud was chosen because it is a small, compact galaxy, and all of it can be captured on a single photographic plate taken with a wide-field camera.

The key to using the period-luminosity relation to calculate distances was first to determine the absolute magnitude of a Cepheid variable in our own galaxy, though by using a different process. In 1918 Shapley used a Hertzsprung-Russell diagram that plots star brightness against surface temperature to determine the absolute magnitudes of eleven Cepheid variable stars in our own galaxy. He was able to calculate the surface temperature by observing the star's spectrum and fitting it onto the diagram. This method works in our own galaxy, where a star will reveal the secret of its spectrum to a telescope. Stars in the distant globular clusters and the Magellanic Clouds are too faint to reveal their spectra.

Once Shapley had the absolute magnitudes to the eleven Cepheids in our own galaxy, he had the key to their distances and the calibration that he needed to determine distances to the globular clusters. He then concluded that the globular clusters were very far—as much as 50,000 light-years—away. He went on to place the center of our own galaxy some 30,000 light-years away.

Until this time, astronomers assumed that our Sun was at the center of the galaxy, a sort of modern-day Ptolemaic viewpoint. Thanks to this work by Leavitt and Shapley, all that changed. The Sun was far from the galactic center, actually in its outskirts. "I stayed with the Cepheids and clusters during those early years at Mount Wilson," he wrote, "until I crashed through on the distances and outlined the structure of the universe. . . . I plotted the globular clusters and looked at what I had. Finally I hit upon using the period-luminosity relation that had been foreshadowed by Miss Henrietta Leavitt at Harvard in a paper published in 1912. Her paper dealt with only twenty-five stars and did not deal with their distances at all. So I went after the distances, and that was helped by Ejnar Hertzsprung's work."[7]

Levy 196   NGC 5694
Globular cluster
First seen: circa 1990
Position (2000.0): α 14 39.6   δ +26 32
Magnitude: 10.2
Distance: 105,000 light-years
Best seen: in spring; dark sky preferred
Tombaugh's globular cluster
Shapley class VII

One of the most recent discoveries of a globular took place in 1932, when the intrepid observer Clyde Tombaugh discovered that the object NGC 5694 was in fact a globular cluster shining at us from the incredibly great distance of 105,000 light-years from the other side of our galaxy.

Examining a pair of plates he had taken on May 12, 1931, centered on Pi Hydrae, in June 1932, he noted in June a 9 magnitude "star" whose edges were sharper than those of stars of similar magnitude. The object was already known as nebulous since William Herschel had first discovered it in May 1784. To Tombaugh's experienced eye, the object looked like a globular star cluster. He went to the standard source, Harlow Shapley's list of globular clusters, and found to his surprise that it was not included. Although the NGC listed the subject as 5694, it did not identify the object as a globular cluster.

Levy 196, NGC 5694, Tombaugh's distant globular.

Two years earlier (see chapter 5), Tombaugh had walked across the hall to the office of his colleague Carl Lampland to inform him of the discovery of a new planet. This time he approached him again to announce, "I think we may have a new globular cluster." After examining the images on Tombaugh's plates, Lampland decided to photograph it with the larger eye of the 42-inch reflector. Thus NGC 5694's true nature as a globular cluster was revealed. This discovery of the ninety-fourth globular cluster was published in August 1932.

Levy 279   NGC 6522 and NGC 6528, Baade's Window
Globular clusters
First seen: May 2004
Position (2000.0) (6522): α 18 03.6   δ −30 02

Position (2000.0) (6528): α 18 04.8 δ −30 03
Magnitudes: 8.6 and 9.5
Distances: 20,000 and 24,000 light-years
Best seen: in summer; needs dark sky
Shapley class VI and V

> Night was only just now falling, but as they approached the
> village it seemed as if a heaven full of countless stars were
> extended before them.
>   —Miguel de Cervantes Saavedra, *Don Quixote* 2, 1615

A heaven full of stars is what one gets when one sees the center
of the Milky Way through a telescope. The stars would be even
thicker if many of them were not obscured by the dust clouds
that lie between us and them. Baade's Window is a break in the
dust clouds. During World War II, while observing at Mount
Wilson (see chapter 8), astronomer Walter Baade identified a
"window" where relatively little dust interferes with our view;
except for a few dust lanes stretching through this window, the
region allows us a rare peek into the galactic center. The region
exposed is only four degrees south of the center and happens
to include these two distant globulars. NGC 6528 may be as
close as 2,000 light-years from the center of the galaxy.

The window has helped us see something quite unexpected: our galaxy may be a barred spiral, its bar not readily visible because it is pointed almost directly at the Sun.

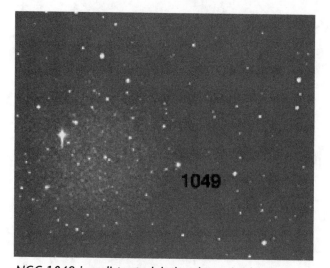

NGC 1049 is a distant globular cluster in the Fornax dwarf galaxy. It is 500,000 light-years away. Photographed by the author using Clyde, a 14-inch Schmidt-Cassegrain telescope with CCD and Starizona Hyperstar. The globular is just above the 1 of the 1049.

CHAPTER 10

FAREWELL TO THE MILKY WAY
THE CLOSEST GALAXIES
DISTANCES: TWO TO FOUR MILLION LIGHT-YEARS AWAY

. . . one of Nature's precious gifts that perchance may come to us but once in a lifetime. . . .

Everybody needs beauty as well as bread, places to play in and pray in, where Nature may heal and cheer and give strength to body and soul alike.

—John Muir, *The Yosemite*, 1912

Now we experience a huge, exponential change in distances as we leave the Milky Way and the mighty globular clusters that surround it. Instead of distances in the tens of thousands of light-years, we now visit objects *millions* of

light-years away. We first become aware of our exit as we see fewer and fewer stars around us. Looking backward, our sky is filled with the spinning spiral of the Milky Way and the star-filled globular clusters around it. Ahead we see the distant spirals of the far-off galaxies. Since all stars belong to galaxies, the space between the galaxies is empty and black. What dominates the sky are the shapes of the twenty-four other galaxies that belong to our galactic family called the Local Group. Two of them are the famous Clouds of Magellan, charted by the great explorer as he voyaged around the world. These galaxies are far from us, but on a cosmic scale they are neighbors of the Milky Way. It is likely that in the distant future, the Milky Way will swallow the smaller one, making its stars part of itself.

*Levy 129   UGC5373, Sextans B
Galaxy in Sextans
Position: α 10 00.0   δ +05 20
Magnitude: 6.6
Distance: 4.3 million light-years
Best seen: in summer and fall; needs dark sky
Classification: Dwarf Irr (see description below)
Local Group Member, HII galaxy

This galaxy looks so much like a faint comet that my heart skipped a couple of beats when I came across it in the course of my comet hunt. I thought it might be a comet, but over time it stayed glued to its home in the sky. This was not a comet, but a galaxy that represents what may be the most common type of galaxy in the Universe: a dwarf irregular galaxy like the Small Magellanic Cloud. The reason we don't see more of them is that they are too far away, and consequently invisible.

I came across this galaxy again on the night of March 7, 1989, just after a partial eclipse of the Sun. It is a slightly oval, dim glow (4.6' x 3.3' in diameter).

Levy 129, Galaxy in Sextans.

THE HUBBLE CLASSIFICATION

Edwin Hubble's classification of galaxies dates back to 1926, when he proposed a "tuning fork" diagram with which he divided galaxies into three categories: elliptical, spiral, and irregular. Spirals and barred spirals are on either side of the fork, and ellipticals, with almost no dust, on the single strand. Ellipticals are classified according to how oval or flattened they appear to be observationally: E0 is a virtually circular galaxy; E7 a flat one. Spirals are classified S or, if barred, SB. An "a" afterward means that the galaxy's arms appear tightly wound, and loose systems get a "c." (The Milky Way is probably intermediate, and since it appears to have a bar it would be an SBb.) A galaxy on the border between spiral and elliptical, a lenticular galaxy, is called S0. Lenticular galaxies are plotted at the beginning of the fork, where it meets the spiral strand. Finally, the irregular galaxies are designated Ir+: they are nearby galaxies resolvable into stars. Ir- galaxies are amorphous, not resolved into stars.

THE LOCAL GROUP

How bright do the galaxies really look from the comfort of our own observing places on Earth? The constellation of Andromeda is the brightest. This galaxy is known as Messier 31, the Andromeda Galaxy, or in the past, the Great Nebula in Andromeda. The Milky Way and Andromeda are the two largest galaxies of a twenty-five-member small clustering called the Local Group. A third galaxy, Messier 33, lies in the constellation of Triangulum. All the other galaxies in the group are small ones. Two of these are so close to the Milky Way that they can readily be called satellites of it. They are the Large and Small Magellanic Clouds. As we have already seen, the LMC has provided scientists with a good deal of information about stars, for in 1987 the brightest supernova in almost four hundred years appeared there.

Levy 80   NGC 185
Galaxy in Cassiopeia
First seen: 1980s
Position: α 00 39.0   δ +48 20
Magnitude: 9.2
Distance: 2 million light-years
Best seen: in fall and winter; requires dark sky
Classification: E0
Almost a quarter of a degree across, this galaxy looks like a faint comet and has fooled me often!

No man reads a book of science from pure inclination.
—James Boswell, *The Life of Samuel Johnson*, 1791

Levy 87   NGC 147
Galaxy in Cassiopeia
First seen: 1980s

Position: α 00 33.2   δ +48 30
Magnitude: 9.3
Distance: 2.3 million light-years
Best seen: in fall and winter; observable in dark sky
Classification: E0

NGC 185 would have fooled me even more often had not its neighbor NGC 147 appeared nearby. Both these dwarf elliptical galaxies are companions of Messier 31.

> If he is to shut himself up for a year to study science, it is better to look out to the fields, than to an opposite wall.
> —Boswell quoting Johnson, *Life of Samuel Johnson*, 1791

Levy 86   NGC 224 M31
The Andromeda Galaxy
First seen: September 9, 1962
Position: α 00 42.7   δ +41 16
Magnitude: 3.4
Distance: 2.5 million light-years
Best seen: in fall and winter; observable in city sky
Classification: Sb

Observing from a clear tenth-century Persian sky, Al-Sufi was the earliest person whose record of the "Great Nebula in Andromeda" survives. But I can imagine a young native American child, standing on the summit of a mountain now called Kitt Peak, in a land now called Arizona, looking up at the sky. The warm season is nearly over. As she looks up, her eye catches a large misty spot high in the sky. Not a star, not a moving cloud, it is something new and different. Perhaps this Native American child is the true discoverer.

She's been dust for thousands of years, but if she could come back and see her mountain today, she'd see a network of great eyes pointing to the stars and learning about the misty spot and others like it. Ten thousand years have passed, but most of what we know about the Andromeda Galaxy we have learned in the last hundred. Indeed, Messier 31 holds the honor of being the first galaxy that exposed its starry interior to the telescopes of Hubble and Baade.

It is often claimed that the Southern Hemisphere has the best objects—47 Tucanae and Omega Centauri as globular clusters, the Southern Cross, Alpha and Beta Centauri, and Eta Carinae. But with galaxies, the Northern Hemisphere wins hands down with objects like the Andromeda Galaxy and the Whirlpool Galaxy M51.

The great galaxy of Andromeda, 2 million light-years out, and our own Milky Way, are the biggest galaxies in the Local Group. As we enter this new star system, we look back at our own galaxy, its 200 billion stars looking like a bright fuzzy spot against the background of space. If we spent enough time here, could we possibly find Earthlike planets with some forms of life?

We have already described (in chapter 8) Walter Baade's successful attempt to relate the nature of M31 to the Milky Way. But what no one in that generation knew was the nature of the two galaxies' motions. These two giant galaxies were probably formed very close together in space and time and are now in a very elongated orbit around each other. They are rushing closer at the phenomenal rate of hundreds of thousands of miles per hour. We now suspect that over the next several million years the two giant galaxies will move closer until they finally collide in 3 billion years.

This first collision will be more of a sideswipe, but the spiral arms of both galaxies will get torn apart as the galaxies, now in strong gravitational contact with each other, begin a dance that brings them into a second, direct collision a few hundred million years later. The process will probably not involve direct encounters with many stars, which are so far apart that they will slide past each other. Even the Sun, which will still be a yellow star, should survive. But the dark clouds of both galaxies will clash, and the tremendous tidal stresses will, over the course of another half-billion years, drastically alter the shapes of both galaxies. Neither galaxy will likely emerge from this titanic collision. Instead, a great new galaxy will form in an elliptical system. Since the two galaxies out of which it will be formed are among the largest spirals, the new elliptical galaxy, born out of the wreckage, will be one of the largest in the Universe. And this grand structure will consist of atoms, molecules, and the dust of what once was our lives and our civilization. That will be way in the future. Let us now look at the past.

OF GALAXY HISTORY: THE EARLY DAYS AT LOWELL OBSERVATORY

Before Vesto M. Slipher, an astronomer at Arizona's Lowell Observatory, did a small experiment almost a century ago, we had no idea about the appearance of objects like NGC 4605. Percival Lowell, the observatory's founder, was primarily interested in Mars, but he suspected that the distant nebulae were planetary systems being formed. To find out, Lowell asked his friend Slipher to photograph these nebulae under the light of a spectroscope. It was a daunting project. Using the observatory's long and majestic 24-inch refractor, Slipher spent as much as two full nights taking a single exposure to gather enough light to reveal the spectrum of a single nebula. Slipher's results were completely unexpected and puzzling: the entire spectra of these nebulae were shifted toward the red end by various amounts. Because of a lack of telescope power, he was unable to interpret these results, and their mystery remained for almost a quarter century.

HOW WE CAME TO UNDERSTAND
THE NATURE OF GALAXIES

More than two centuries ago, Immanuel Kant suggested that we lived in a system of stars and that there were other systems like ours. In the years around 1800, William Herschel compiled a catalog of some twenty-five hundred objects that he saw in his survey of the sky—most of them turned out to be galaxies. In 1850 the German astronomer Alexander von Humboldt suggested the name "island universe" for each of these distant objects, an idea far ahead of its time. We don't use the term anymore, but the idea of each galaxy being its own, self-contained island has a certain romance to it. Instead, until recently we used the terms extragalactic nebulae, or anagalactic nebulae, to describe these distant galaxies.

In 1888 John Dreyer's *New General Catalog*, or NGC , described almost eight thousand "nebulae"; in 1908, two additional Index Catalogs (IC) appeared, raising the total number of objects to thirteen thousand. Almost 90 percent of these fuzzy objects are distant galaxies. That number is paltry compared with what we have now seen and what more modern catalogs include: galaxies in space virtually without number. And if our findings of small, irregular galaxies close to the Milky Way are any indication, the number of galaxies we have seen must be multiplied many times if we include many smaller systems too small and too distant to be seen with our telescopes.

Levy 91   NGC 598 M33
Galaxy in Triangulum
First seen: July 31, 1964
Position: α 01 33.9   δ +30 39
Magnitude: 5.7
Distance: 2.7 million light-years

Best seen: in summer and fall; observable in dark sky
Classification: Sc

Levy 91, M33. Triangulum Galaxy.

The Pinwheel Galaxy, M33, is a test of a dark night. If you can see this almost 6 magnitude galaxy, whose light is spread across more than a degree (71') of sky, without any optical aid, then you know you have a very good, world-class sky from which to observe.

Discovered by Charles Messier in 1764, this beautiful spiral galaxy was the subject of a search for extraterrestrial life. A large galaxy viewed face-on might display signs of a galaxy-wide communication network if it were home to a galactic civilization. But no evidence of such an advanced network has ever been found.

SPIRAL GALAXIES

Spiral galaxies, of which M33 and M31 are striking examples, are among the most graceful in the sky. They are also the most common of the galaxies we can observe; at least four-fifths of the galaxies we know of are spiral in form. Their spiral arms stretching out dramatically, these galaxies offer many hours of pleasant viewing through a telescope. According to Edwin Hubble's classification, spirals range, as noted, from being very tightly wound (Sa), to moderately wound (Sb) (which is what we used to believe our galaxy to be), to very loose, with arms spread over a wide area (Sc). Some galaxies have a central bar in addition to a fattened nucleus, with the arms spreading out from the ends of the bar. New evidence suggests that the center of the Milky Way might actually be a bar and that our galaxy is a barred spiral.

HOW GALAXIES CAME TO BE

The early Universe, according to the Big Bang theory, was a soup consisting of hot gas and unseen matter. Although this broth was quite homogeneous in structure, some parts of it were denser than others. The dense regions had slightly more gravity than the regions surrounding them, so that eventually these areas began to circulate, pulling in the matter around them to form massive protogalaxies. Matter within these pro-togalaxies began to contract toward their centers and spin more quickly, just like a figure skater spins faster as she brings her arms closer to her body. After most of the gas settled into a galaxy's center and surrounding disk, it began to condense fur-ther to begin the process of star formation. The gas was spin-ning so rapidly that it could not fall to the center; conse-quently, much of it turned into stars in a galaxy's disk and spiral arms. The faster the gas was spinning, the larger in area the resulting galaxy became. A protogalaxy that spun rapidly formed a galaxy of larger area than one with the same amount of gas spinning more slowly. Within the relatively brief timescale of perhaps a few hundreds of millions of years, galaxies began to form in great numbers and in vast groupings. As we leave our own Local Group and travel farther out, we will meet other fabulous "island universes."

CHAPTER II

WHERE SPACE IS FILLED WITH GALAXIES
DISTANCES: FOUR TO SIXTY MILLION LIGHT-YEARS AWAY

. . . the night was springlike, still and mild,
the stars thick-sown in a faintly hazy sky.
—Henry Beston,
The Outermost House, 1928

As we dart away from our home galaxy at many times the speed of light to get to the next cluster of galaxies in the constellations of Virgo and Coma Berenices, we travel some 50 million light-years from home. It is impossible to travel at the speed of light, but even if we could, it would take us 50 million years to reach this nearest edge of the cluster. At the speed of thought, however, we can. As we reach the galaxies of Virgo and Coma Berenices, we realize that our Local Group is gravitationally bound to this cluster—thousands of galaxies are sharing the same part of space, sharing the same destiny, all part of a big array called the Local Supercluster. Its largest galaxies are elliptical monsters, others are great spirals like our Milky Way, and still others are irregular.

Levy 2   NGC 5457 M101
Pinwheel Galaxy in Ursa Major
First seen: July 4, 1966
Position (2000.0): α 14 03.2   δ +54 21
Magnitude: 7.9
Distance: 17.5 million light-years
Best seen: in spring, summer, and fall; observable in dark sky
Classification: Sc

In chapter 3 I described, as an example of how I record my observing sessions, the time I first saw M101. Independence Day 1966 was clear as a bell over the Adirondack Science Camp. With the onset of night I set up Pegasus and began an evening of comet hunting. A fellow counselor came by and asked why I wasn't at the festivities. When I said that I wanted to get as much observing in as I could before moonrise, he countered that I was expected to be supervising the campers. "But the director said I can observe any night of camp!" I argued. He returned to the kids; I returned to the scope, anxious to observe as much as possible before the Moon rose.

I don't know if that was the right thing to do; after all, I thought the children at this particular camp did not require much supervision. One child, Alex Scheeline, went on to become a professor of chemistry at the University of Illinois, Champaigne/Urbana; another, Steve Ashe, would become a first-rate artist and astronomer; Andy Bauman would become a physi-

Levy 2, M101, Pinwheel galaxy.

cian; and David Larach would earn both a PhD in cardiovascular pharmacology and an MD. The campers were more than campers; they were the friends of my youth, and I am still in touch with all four of them, except Steve who died from cancer in 2003. Not only are we still in touch, but we have gotten together to begin, at the same campsite, an annual Adirondack Astronomy Retreat where, for a few days each summer, we get together to recharge our batteries. We're also building the observatory that I described in chapter 1 at that site.

So it was during this stealth observing session that I encountered Messer 101. It was so large and diffuse, and its beautiful spiral structure was so easy to spot, that I knew it couldn't be a comet. A spectacular spiral galaxy, M101 would be chosen a quarter-century later as one of the first targets for the Hubble Space Telescope. But on the night of July 4, 1966, a telescope in space was only a dream. As I watched, I heard the loud boom of a fireworks display over the camp. That's great I thought. They're seeing their fireworks, and through the telescope, looking at M101, I'm seeing mine.

Levy 7   NGC 5055 M63
Sunflower Galaxy in Canes Venatici
First seen: April 18, 1965
Position (2000.0): α 13 15.8   δ +42 02
Magnitude: 8.6
Distance: 23.5 million light-years
Best seen: in winter and spring; observable in suburban or dark sky
Classification: Sb

This well-named Sunflower Galaxy has a dense core, like the head of a sunflower, and its arms resemble a complex series of petals that overlap each other. It is resemble in that it seems to have two sets of arms (at least as seen in photographs), one tightly wound about the center, the other spreading out loosely at great distances from the core.

Levy 9   NGC 2403
Galaxy in Camelopardalis
First seen: July 16, 1966
Position (2000.0): α 07 36.9 δ +65 36
Magnitude: 8.4
Distance: 14 million light-years
Best seen: in spring; observable in dark sky
Classification: Sc

> And what's the sky? Air and scattered light; but also a symbol of that boundless and (excuse the metaphor) pregnant emptiness out of which everything, the living and the inanimate, the puppet makers and their divine marionettes, emerge into the universe we know—or rather that we think we know.
> —Aldous Huxley, *Island*, 1962

A beautiful spiral, looking like a faint smudge of light at low power, it reveals itself into a beautifully complex spiral galaxy when you increase the magnification. In August 2004 a 12 magnitude supernova flared up in this galaxy. It is especially satisfying to see a single star whose cataclysmic demise occurred long before the dawn of humanity.

Levy 11   NGC 4605
Galaxy in Ursa Major
First seen: August 5, 1966
Position (2000.0): α 12 40.0 δ +61 36

Magnitude: 10.3
Distance: 16 million light-years
Best seen: in spring; observable in dark sky
Classification: SBc pec

A strange barred spiral galaxy, almost edge-on from our point of view, NGC 4605 clearly shows a bar across its center oriented southeast to northwest. It is very close—only three-fourths of a degree southeast—from the first Hubble Space Telescope deep field that became so famous after the image was released early in 1996.

Levy 15   NGC 4826 M64
The Black Eye Galaxy in Camelopardalis
First seen: July 15, 1966
Position (2000.0): α 12 56.7   δ +21 41
Magnitude: 8.5
Distance: 13.5 million light-years
Best seen: in winter and spring; requires suburban or dark sky
Classification: Sb

This galaxy really does look like some cosmic force punched it. Like a black eye, a long black cloud stretches some 40,000 light-years across its face. Its dust lane is raw material that someday will be part of stars and planets, just as long-gone dust clouds within our own galaxy are now a part of you, dear reader, and me. Looking at dust-laden M64 reminds

Levy 15, M64, Black Eye galaxy.
Tim Hunter photo.

us of this, and it also brings to mind a debate that took place many years ago about where the distant galaxies stood in the overall structure of the universe.

THE GREAT DEBATE

April 26, 1920, was an unforgettable day in astronomy, a day focused on the National Academy of Sciences in Washington, where Harlow Shapley and Heber Curtis were to come together to discuss the state of the Universe. What was at stake that day was our understanding of its size and nature. Who would win: Shapley or Curtis? The answer: neither. The winner was a young astronomer named Edwin Hubble, who, at the time, was nowhere near Washington.

What were the spiral-shaped nebulae, and how far away were they? From his studies of the variable stars, Shapley knew that our galaxy was some ten times bigger than scientists had previously thought. From this, he thought that the spirals could not lie very far outside our galaxy. Curtis, using older reasoning, thought that our galaxy was small and that these remote spiral-shaped fuzzy patches were comparable in size and nature to our own galaxy.

Although Shapley's reasoning about the size of our galaxy was correct, he was wrong about the nature and distance of the spiral nebulae. Curtis was right about the spirals, but for the wrong reason! Using the observational data of the time, both scientists did the best they could. With his telescopes, Shapley could measure distances no farther than the globular clusters and the Magellanic clouds, which are really at the outskirts of our own galaxy. Just four years later, Hubble used Shapley's own yardstick, the variable stars, to prove Shapley wrong in the great debate. Using the newly opened 100-inch telescope on Mount Wilson, Hubble found stars within these spirals. These stars

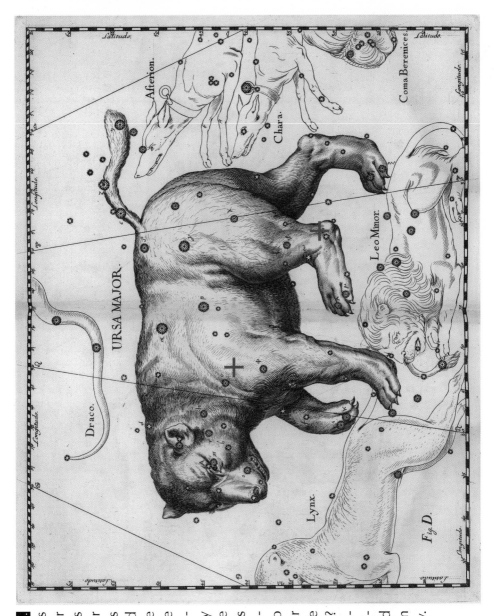

1. In 1690 Johannes Hevelius published his famous star atlas *Uranographia*. This printout of the Ursa Major page contains two objects that would have fascinated Hevelius. The cross on the left marks the approximate position of the gravitationally lensed quasar Levy 337 Q0957+56A/B. The cross on the right covers Levy 297 47 Ursae Majoris, a star with at least two planets in near-circular orbits. Could there be life on worlds circling this star? This image was reproduced from the first edition, housed in the Harold B. Lee Library of Brigham Young University.

2. The Pleiades (Levy 336), photographed by Jack Newton from Arizona Sky Village. Williams f/6.5 camera, 80-minute exposure.

3. Nightfall, a precious time for observers. Wendee Walllach-Levy examines Venus and Jupiter as she prepares for an evening of observing.

4. The Cone Nebula (see Levy 159), photographed by Dean Koenig of Stari-zona, using a 4.5-inch telescope.

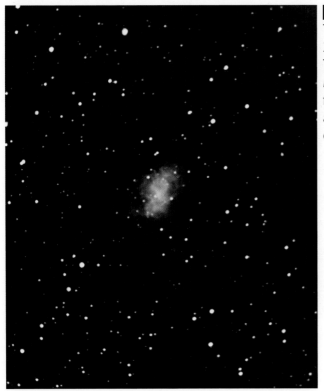

5.
The Crab Nebula (Levy 33), photographed by Tim Hunter using a Meade LX200 12-inch f/6.5 telescope and an Apogee AP 7 CCD camera.

6. May 2004; David Levy searching for comets as the center of the Milky Way galaxy rises in the southeast. Photograph by Tim Hunter.

7. Globular cluster M80 (Levy 117), photographed by the author, using Esther, a 10-inch LX 200 telescope, and a Meade DSI CCD camera.

8. Globular cluster M15 (Levy 12), photographed by Tim Hunter with his Meade LX200.

9. Andromeda Galaxy (Levy 86), photographed by Jack Newton. One-hour combined exposure at f/6.

10. Ever since the third earl of Rosse drew the structure of the Whirlpool Nebula in early 1845, Messier 51 (see Levy 87) has been admired as a beautiful structure. This Hubble Space Telescope picture, released in spring 2005, is actually a composite of many images. Never before has the Whirlpool appeared this way, as one of the truly magnificent spiral galaxies in our night sky.

11. The author, searching for comets through Miranda. Photograph by Wendee Wallach-Levy.

12. NGC 6946 (Levy 250), photographed by Wendee Wallach-Levy and Adam Block, and David Levy using a 20-inch RC Optical Telescope at Kitt Peak National Observatory.

were so far away that the spiral nebulae in which they lived had to be distant galaxies. The Hubble Space Telescope has found these variable stars in other galaxies. In this way, the work Shapley did three-quarters of a century ago continues on and on. As long as telescopes get bigger and better, we will be able to measure distances farther and farther out.

While Vesto Slipher used a 24-inch refractor at Lowell Observatory in northern Arizona to show us that a mystery existed, Edwin Hubble had the great 100-inch reflector at Mount Wilson at his disposal to open that window and see clearly through it.

Levy 16   NGC 3627 M66
Galaxy in Leo
First seen: March 11, 1966
Position (2000.0): α 11 20.2 δ +12 59
Magnitude: 9.0
Distance: 21.5 million light-years
Best seen: in spring; observable in dark sky
Classification: Sb

Messier 66 is part of a really delightful trio of galaxies, of which M65 and NGC 3628 are the other members. While M65 is almost edge-on in appearance, M66 is angled so that we see more of its face, including one spiral arm that hangs more limply than the other, as if the galaxy had suffered some cosmic fall that injured its shoulder.

ORIENTATION OF GALAXIES

If you really want proof that the Universe is a three-dimensional place, look at different galaxies. Some, like M66, we see at a sharp angle, while others, like M65, are shown more face-on. But M101 is portrayed to us exactly face-on, so we can admire its structure in glorious detail.

Levy 22   NGC 3034 M82
Cigar Galaxy in Ursa Major
First seen: September 7, 1964
Position (2000.0): α 09 55.8   δ +69 41
Magnitude: 8.4
Distance: 12.2 million light-years
Best seen: in spring; observable in suburban sky
Classification: Irregular

Levy 22, M82, a starburst galaxy.
Tim Hunter photo.

Of all the galaxies that are bright enough to be seen from a suburban sky, Messier 82 is the most peculiar. Located in Ursa Major next to its neighbor M81, this galaxy totally lacks the spiral symmetry of its neighbor. It looks as if it had suffered some gigantic cataclysm at the hands of M81. It and M81 are also the best galaxies to spot from a suburban location during the Northern Hemisphere's autumn months.

What really happened to M82? Forty years ago I learned that M82 was an exploding galaxy. Twenty years later, I learned that, instead, it was undergoing a burst of star formation. The present theory explains that the star formation was set off by a close encounter with M81. As M81 raced past M82 some 40 million years ago, a gravitational shock wave plummeted though the smaller galaxy, causing its supply of interstellar hydrogen to start coalescing and forming new stars. Then, as the two galaxies receded, a second burst of star formation began. We may be witnessing the tail end of that formation today.

Levy 24   NGC 3031 M81
Bode's Galaxy in Ursa Major
First seen: September 7, 1964
Position (2000.0): α 09 55.6   δ +69 04
Magnitude: 6.9
Distance: 11.8 million light-years
Best seen: in summer and fall; needs dark sky
Classification: Sab

> The sky was without a cloud; and the dawning mystery of moonlight began to tremble already in the region of the eastern heaven.
>
> —Wilkie Collins, *The Woman in White*, 1860

For a few nights after the full Moon, the early evenings can be dark and clear for a short time before the Moon rises in the "region of the eastern heaven." One can try to take advantage of that brief hour or two of pre-Moon darkness, especially on spring evenings, to peer 12 million light-years into space to find M81, "a massive, beautiful galaxy," as I wrote in one of my early impressions of it.

HOW FAR CAN WE SEE?

This child's question begs for an answer, and it gets one in our discussion of the big galaxy called Messier 81. Yes we can see down the block, and we can see the distant church steeple at the end of the road. From a plane we can see several miles down to the ground, and on the ground, we can gaze at the horizon, which, if flat, can be 30 miles away. But Barbra Streisand starred in a movie whose title was right: on a clear day you *can* see forever. You can see the Moon at 240,000 miles away, and even without a telescope, from a fairly dark suburban sky, you can make out Messier 31—the Andromeda Galaxy.

The farthest galaxy I have ever seen is Messier 81. It is not easy for me—I can accomplish this feat only when the sky is dark and sparkling. But if the distance I have used, from the Tully Database of Galaxies, is correct, then I can see almost 12 million light-years into space—and all without any optical aid whatsoever.

In late summer 1989, Steve O'Meara observed Comet Okazaki-Levy-Rudenko using a large refractor at Amherst College in Massachusetts. He said, "The comet floated past M81 and M82 like a specter in a long white gown, and I felt like I was transported into Wilkie Collins's novel *The Woman in White*, just when Walter Hartright first encounted the specter: 'I was far too seriously startled by the suddenness with which this extraordinary apparition stood before me, in the dead of night and in that lonely place, to ask what she wanted.'"[1]

O'Meara's comparing the mysterious woman in white to the apparition of a comet as it slides gracefully through the night sky past two distant galaxies is a metaphor that drew my attention to Collins's tale. As the reference at the start of this little section on M81 shows, the novel alludes to the mystery of the night sky. It also bemoans how little interest many people have in the natural world: "At any time, and under any circumstances of human interest, it is not strange to see how little real hold the objects of the natural world amid which we live can gain on our hearts and minds." It is my desire that this guide will provide the tools that will allow natural wonders like M81 to gain that hold in your eyes, your hearts, and your minds.

Levy 26   NGC 4258 M106
Galaxy in Canes Venatici
First seen: July 7, 1966
Position (2000.0): α 12 19.0   δ +47 18
Magnitude: 8.3
Distance: 22 million light-years
Best seen: in winter and spring; observable in suburban or dark sky
Classification: Sb

A spiral galaxy trying to tell us something about its past, Messier 106 has an odd shape for a spiral. When observed under high magnifcation, the outer arms in particular seem to stretch far and away from the galaxy's center. The structure is graceful, but the dark blotches of material within them are not, each one a rich area of activity preceding the formation of new stars. It is a "long object" through Pegasus, as I wrote in my first observation in 1966, "easy but conditions were not perfect— about as difficult as M101."

*Levy 45   NGC 2188
Galaxy in Columba
First seen: December 17, 2001
Position (2000.0): α 06 10.0   δ –34 06
Magnitude: 11.8
Distance: 32.1 million light-years
Best seen: in summer and fall; needs dark sky
Classification: Sd

I call this object the "Columba comet tail" because it looks like the tail of a comet, especially when seen low in the sky. It really fooled me at first. I thought it was a comet, but as often happens, it stayed plastered to its place in the sky until I identified it. NGC 2188 is also a starburst galaxy, undergoing a phase of active formation of new stars.

Levy 52   NGC 3115
Spindle Galaxy
Galaxy in Sextans
First seen: 1999
Position (2000.0): α 10 05.2   δ −07 43
Magnitude: 8.9
Distance: 22 million light-years
Best seen: in spring; observable in city sky
Classification: S0, Lenticular

*Levy 52, NGC 3115,
the Spindle Galaxy.
Tim Hunter photo.*

This galaxy's condensed nature makes it a nearly ideal target for observers under light-polluted skies. This was the first object I saw using my new Meade ETX telescope with its Autostar computer system. I had never observed this way before. After asking the Autostar to take me on a tour, it recommended the Spindle Galaxy as its first object. Since I was not too familiar with the Spindle, I accepted the suggestion by pressing the "go to" button, and the telescope moved all by itself to the galaxy, which was not in the middle of the field of view, but it was in the field. As I looked at the faint, long fuzzy light in the field of the telescope, I pressed the "?" button on the tele-scope's hand paddle, asking the telescope to tell me what it knew about this object. "This extremely flattened, elliptical galaxy," the telescope explained, "is approximately 27 million light-years away [different sources give different distances for galaxies], and 30,000 light-years in diameter." (It is perhaps more accurately called a lenticular galaxy, on the edge between spirals and ellipticals.) Through a small telescope, look for a tiny elongated patch of light. "It is thought that a black hole whose mass equals two million stars like our Sun, lies at the center of this galaxy." Using this telescope was like having an expert at your side; I was amazed, frankly, that my telescope could be the tour guide that I wished I'd had when I started astronomy years ago.

All this technology depends on the user taking the time to set up the telescope properly in the first place. Once that is accomplished, the telescope is capable of opening whole new vistas, even for experienced observers. These telescopes are revolutionizing the way we look at the stars.

Levy 54   NGC 5128
Galaxy in Centaurus
First seen: December 19, 2001
Position (2000.0): α 13 25.5   δ –43 01
Magnitude: 6.7
Distance: 19 million light-years
Best seen: in spring; observable in suburban sky
Classification: S0, peculiar

Also known as Centaurus A, this galaxy's peculiarity is obvious even with a small telescope. On the morning of December 19, 2001, while comet hunting with Pegasus, I picked up this galaxy as it rose, touching the distant mountain peaks of the Empire Mountains at the southeastern horizon. With the lack of definition that happens when a galaxy, or any other object, is rising, the galaxy looked so much like a comet that I was certain I had one for a short time. The fact that Pegasus's encoder system had skipped a beat and wasn't recording correctly where the telescope was pointing added to the confusion. But the "thrill-letdown" routine that thwarts comet hunters so often struck again that morning, and all I had to add to my list for my efforts was this resplendent galaxy. Of course, in hindsight I am really glad I had this experience. I learned two

Levy 54, Centaurus A galaxy.

things: (1) always check the accuracy of the encoders before getting excited, and (2) don't trust the appearance of an object when it's just a degree or so from the horizon.

What I discovered that night was not a comet but a unique view of a unique object. As a galaxy, Centaurus A is special. It seems almost bisected by a complex dust lane. Discovered by James Dunlop from New South Wales, he added it to his catalog in 1827, calling it a double nebula. In 1949 early radio telescope observers related the strong source they called Centaurus A as identical with NGC 5128. It is one of the strongest radio sources in the sky.

Only recently has the Hubble Space Telescope managed to punch through the enigma of this strange and mighty behemoth of stars, gas, and dust. NGC 5128 is an elliptical galaxy that has been cannibalizing a spiral galaxy for the past several hundred million years. Together the systems contain more than a trillion times the mass of the Sun and are 80,000 light-years wide. The galactic meal should be fully digested within another few hundred million years. It is likely that a supermassive black hole lies at its core.

Finally, NGC 5128 is a Seyfert galaxy with a starlike nucleus. Sometimes the nuclei are slightly variable, in leisurely fashion with ranges of up to three magnitudes over periods of months or years, although much faster changes of a fraction of a magnitude have been observed over a period of days. M 77 and NGC 5128 are examples of this class of galaxy.

When Carl K. Seyfert described these galaxies in 1943, neither he nor anyone else could foresee their future importance. In June 1960 he died in an auto accident in Tennessee while driving home from his second job as a broadcast weatherman. The obituary in *Sky & Telescope* from that August summarized his rich career without mention of the Seyfert galaxies.

Levy 63P   NGC 4656
"Hummingbird" Galaxy in Canes Venatici
First seen: July 6, 2002
Position (2000.0): α 12 44.0   δ +32 10
Magnitude: 10.4
Distance: 29 million light-years
Best seen: in spring and summer; observable best in dark sky
Classification: SB irregular

The veins of leaves dark in the moonlight.
—Sheila Watson, *The Double Hook*, 1966

Levy 63P, The Hummingbird.

This galaxy looks like a small Hummingbird on our photographic films that captured it in 2002. It is, however, more commonly called the Hockey Stick (or the Hook) because of the strange distortion it has at the end of one of its spiral arms. The cause of this yanking away of the end of an arm is probably a close encounter millions of years ago with its neighbor, NGC 4631.

Levy 64   NGC 404
Galaxy in Andromeda
First seen: July 4, 2002
Position (2000.0): α 01 09.4   δ +35 43
Magnitude: 10.3
Distance: 8 million light-years
Best seen: in fall and winter; needs dark sky
Classification: S0

One night in the summer of 1988, my friend Jim Scotti called to tell me that he thought there was a comet near the bright star Mirach, or Beta Andromedae. A few years later my friend Dean Koenig had the same experience, this time while observing visually from Kitt Peak. It wasn't a comet; it was NGC 404, a trickster so buried in the glare of the star that some atlases don't show it. When two experienced and talented observers, using different methods, both run into NGC 404 this way, it's time to put a red flag on it! It's not a comet. It is a galaxy 8 million light-years away that happens to be near a star 200 light-years away. It also has the distinction of being about midway between two great and closer galaxies, M31 and M33.

Levy 65   NGC 4374 M84
Galaxy in Virgo
First seen: July 15, 1966
Position (2000.0): α 12 25.1 δ +12 53
Magnitude: 9.1
Distance: 51 million light-years
Best seen: in spring; observable in suburban sky
Classification: S0 (between a spiral and an elliptical)

"Avoid the Milky Way," comet hunter Robert Burnham told me during the afternoon of June 6, 1967, as he guided us around the Lowell Observatory, "and the Virgo group. And if you never give up, someday you will find a comet." I've never forgotten those words, and for years I studiously followed them. When I switched to the larger 16-inch, Miranda, in 1982, I started including the Milky Way in my search because the larger aperture made it easy to search through its star-filled regions. By the mid-1980s I had come to tell the structural difference between many of these galaxies and a comet, so I became less averse to the region. But when I added an encoder system to my telescope, I no longer avoided the realm of the galaxies. With encoders, I didn't have to spend precious dark time identifying each of the galaxies; the system did it for me instantly.

Messiers 84, 86, and their smaller ilk are near the heart of the Virgo cluster. Their field is stunning. With two other smaller galaxies, they form an almost literal constellation of Sagitta the arrow, but it is formed out of galaxies, not stars. It is also a part of Markarian's galaxy chain (about which we'll hear more later).

Our Local Group belongs to a throng of clusters of galaxies together called the Local Supercluster. The member clusters are in the constellations of Ursa Major, Coma Berenices, Leo, and Virgo. Messiers 84 and 86 are in the middle of the Virgo cluster. The gravitational pull of these galaxies is affecting the Milky Way; our home galaxy is part of the extended family, the Local Supercluster, centered on the Virgo group. Though I have never found a comet here, it is always a special experience to move my telescope, field after field, through the maze of galaxies that lurks behind the stars of Virgo.

Levy 66   NGC 4406 M86
Galaxy in Virgo
First seen: July 15, 1966
Position (2000.0): α 12 26.2   δ +12 57
Magnitude: 8.9
Distance: 50 million light-years
Best seen: in spring; observable in suburban sky
Classification: E0

> He compared reason to the sun, of which the light is constant, uniform, and lasting; and fancy to a meteor, of bright but transitory lustre, irregular in its motion, and delusive in its direction.
>
> —Samuel Johnson, *Rasselas*, 1759

Even though M84 and M86 are companion galaxies, they appear quite different through a telescope. M86 is a much larger galaxy, shining at us from near the center of the cloud. "Brighter and Bigger than M84" was my initial comment in 1966. M86 lacks the dense core of its neighbor, but makes up for that by the grace of its oval structure.

Levy 67   NGC 4594 M104
Galaxy in southern Virgo
First seen: May 5, 1967
Position (2000.0): α 12 40.0   δ −11 37
Magnitude: 8.0
Distance: 35 million light-years
Best seen: in spring; observable in suburban sky
Classification: Sa

Levy 67, M104,
Sombrero galaxy.
Tim Hunter photo.

"Worthy of name Sombrero," I wrote when I saw the beautiful galaxy with a dust lane. It lies in a rich region of sky just over five degrees—the separation between the two stars at the end of the Big Dipper's Bowl—from Eta Corvi. Better still, take the stars at the southwest and northeast ends of the quadrilateral of Corvus the Crow, and extend the line an approximately equal distance toward the northeast. M104 is just west of the end of the line. One of the brighter galaxies, the Sombrero with its thick lane of dust is well worth a look on a clear spring night.

Levy 68   NGC 1023
Very elongated galaxy in Perseus
First seen: July 4, 2002

Position (2000.0): α 02 40.4   δ +39 04
Magnitude: 9.5
Distance: 30 million light-years
Best seen: in fall; needs dark sky
Classification: S0

NGC 1023 is a really magnificent spiral galaxy, especially at medium power. Because its field is near the Perseus Milky Way, the stars surrounding it are many and striking. On the evening of January 31, 2005, I saw two faint stars on either side of this galaxy, as other observers have reported. The galaxy is also surrounded by a triangle of brighter stars.

Levy 74   NGC 3623 M65
Galaxy in Leo
First seen: March 11, 1966

Levy 87, M51, Whirlpool galaxy. Tim Hunter photo.

Position (2000.0): α 11 18.9   δ +13 05
Magnitude: 9.3
Distance: 30 million light-years
Best seen: in spring; observable in suburban sky
Classification: Sab

Both M65 and M66 were discovered by Messier's colleague and rival, Pierre Méchain, in 1780. When Messier heard of the finds, he inspected them himself, and he subsequently added them to his catalog. Because they form such an interesting pair of galaxies on spring nights, I have added them to my list as well.

Levy 87   NGC 5194 M51
Whirlpool Galaxy in Canes Venatici
First seen: April 18, 1965
Position (2000.0): α 13 29.9   δ +47 12
Magnitude: 8.4
Distance: 26 million light-years
Best seen: in winter and spring; observable in city sky
Classification: Sbc

> It is now three o'clock in the morning, and I have just completed an observing session that couldn't have lasted half an hour, yet turned out to be a fine communion with a part of Nature which I have always loved, but in these last months have ignored. I realize tonight that it does not matter whether I hunt for comets, or obtain magnitude estimates of variable stars, or stay out all night. The good observing session means a private feeling of a successful rendezvous with Vega or Jupiter, as in tonight's case, or Saturn and Sirius and Canopus on another night.
>
> —my own daily journal, May 22, 1973

Levy 92, NGC 253.
Caroline Herschel's galaxy.

Special experiences like this are what make observing special. In June 1997 Wendee had such an experience. "This is the first time I saw a galaxy come to life," is how she described her first view of the great Whirlpool through the 61-inch telescope atop Mt. Bigelow, Arizona. The galaxy seemed to have no end as it spiraled outward from its center. It was a staggering sight. It was the first time Wendee actually saw the separate arms unfolding from the galaxy's center, and we were both entranced.

The Whirlpool is the best example of a face-on spiral. But it's more than a single galaxy; its companion NGC 5195 appears to hang onto the edge of one of the Whirlpool's arms as it orbits the larger galaxy. Imagine what the Whirlpool would look like for someone living on a planet in NGC 5195; it would take up most of the sky, its graceful structure ruling the night.

Messier discovered this galaxy on October 13, 1773. We now know that the Whirlpool is some 50,000 light-years across, and it shines with the intensity of ten billion suns.[2] It can be spotted as a diffuse patch of light through small telescopes, but, through large telescopes, the vastness of the spiral structure is spectacular.

Levy 92   NGC 253
Caroline Herschel's galaxy in Sculptor, Silver Coin, Great Sculptor
First seen: October 1979
Position (2000.0): α 00 47.6 δ–25 17
Magnitude: 7.6

Levy 144, M88. Photo by David Levy and Steve Larson through the Kuiper 61-inch telescope, Mt. Bigelow, Arizona.

Distance: 8.8 million light-years
Best seen: in fall; requires a dark sky
Caroline Herschel, 1783
Classification: Scd

What a fabulous galaxy this is! Too far south to appreciate properly until I relocated to Arizona, I found it just a few weeks after I settled in to my new surroundings. It appeared at a sharp angle, almost edge-on. It has some massive dust lanes, and if it were seen face-on, its arms would be glorious. Caroline Herschel found the galaxy with a small Newtonian telescope. From the latitude of Slough, in England, it was pretty close to the southern horizon. In fact, it is surprising that she found it. NGC 253 is the brightest member of the Sculptor Group, the closest one to our Local Group. It is a starburst galaxy, which means that its central region has undergone recent violent episodes of star formation.

MEETING CAROLINE
AND WILLIAM HERSCHEL

As Tricia Brown of the Bristol Astronomical Society drove me to a lecture in Bristol, England, in September 2003, we couldn't resist stopping by 19 New King Street in nearby Bath, the home of William and Caroline Herschel. I felt as though I was visiting old friends. Here I was ushered into the Herschel house, where I imagined hearing William talk about his planet, Uranus. I imagined him sharing with me the congratulatory letter he got from comet hunter Messier, and then Caroline musing how searching for comets might be a worthwhile pastime. Turning toward his sister, William would smile, "Lina, yes! I think you could find a comet someday." (Caroline found the first of her eight comets in 1786, after they had moved near Windsor Castle.)

It is extraordinary how a visit to some key places can actually be a substitute for a real meeting with a person one will never actually meet. But the strongest connection came not inside the house but outdoors in the garden, where Herschel had set up his 6-inch reflector in 1781 to conduct his sky survey, and where, on March 13, he discovered the solar system's seventh planet. As I looked up in the northwest, I mentally placed H Geminorum in the sky and imagined hearing Herschel's words: "While I was examining the small stars in the neighborhood of H Geminorum, I perceived one that appeared visibly larger than the rest; being struck with its uncommon magnitude I compared it to H Geminorum and the small star in the quartile between Auriga and Gemini, and finding it so much larger than either of them, suspected it to be a comet."

I left the Herschel museum struck by this experience. I had met and known Clyde Tombaugh, and now I felt I knew the Herschels. Just two years after Herschel's find, Caroline would discover NGC 253 as it hung low in her southern sky. Although she didn't find it from Bath, I felt that the house gave me a connection with two people I had always admired.

Levy 144   NGC 4501 M88
Galaxy in Virgo
First seen: July 15, 1966
Position (2000.0): α 12 32.0   δ+14 25
Magnitude: 9.5
Distance: 54 million light-years
Best seen: in spring; needs dark sky
Classification: Sbc

Between August 1985 and summer 1989, I worked with Steve Larson as part of the International Halley Watch, a worldwide effort to coordinate amateur and professional observations of Halley's comet. Our division centered on near-nucleus studies. Each month during this time, we would spend two nights at the 61-inch telescope on Mt. Bigelow imaging comets, particularly Halley's. On one of these nights, Steve Larson asked me to select a galaxy that would fit nicely into the small field of the 61-inch telescope we used, coupled together with the International Halley Watch CCD. I selected Messier 88, whose dimenions of 6.9 by 3.7 arcminutes would fit perfectly into the field. The resulting picture shows a graceful spiral galaxy.

Messier 88, photographed from Mt. Bigelow, Arizona, using the 61-inch telescope by David Levy and Steve Larson.

*Levy 145   NGC 4473

Galaxy in Coma Berenices, elongated with a bright core

First seen: October 18, 2002

Position (2000.0): α 12 29.8   δ +13 26

Magnitude: 10.2

Distance: 58 million light-years

Best seen: in spring; requires dark sky

Classification: E0

As diffuse as this galaxy is, it can easily fool a comet chaser. Observationally, it is called a "low surface brightness" galaxy, which means that even in a dark sky, it isn't much brighter than the surrounding sky background. NGC 4473 is a member of a string of galaxies called "Markarian's Galaxy Chain," which includes M84 and M86, and continues past NGC 4473 to NGC 4478.

*Levy 173   NGC 4548 M91
Galaxy in Coma Berenices with bright core
First seen: February 17, 1983
Position (2000.0): α 12 35.4   δ +14 30
Magnitude: 10.2
Distance: 55 million light-years
Best seen: in spring; observable in dark sky
Classification: SBb

For many years Messier aficionados considered M91 a "missing Messier," possibly a comet that got away, possibly the same as M58. But a small letter in the depths of *Sky & Telescope* cleared up the mystery. Its author, William C. Williams, suggested in December 1969 that M91 is actually a misplotted NGC 4548. Williams's logic was simple. Because there were no good reference stars in the galaxy's vicinity, Messier used M89, a nearby galaxy, to derive positions for his new find. Williams then suggested that Messier made a mistake. When he calculated the position of NGC 4548, he applied the offset positions to nearby M58, not NGC 4548. If this is so, then the resulting position is pretty close to NGC 4548. As odd as this might first appear, the deep sky community has generally accepted this explanation, and NGC 4548 is now considered to be M91.[3]

*Levy 174   NGC 4649 M60
Galaxy in Virgo
First seen: April 15, 1966

Position (2000.0): α 12 43.7   δ +11 33

Magnitude: 8.8

Distance: 59 million light-years

Best seen: in summer and fall; observable in suburban sky

Classification: E

One of the largest ellipticals, M60 is more than 100,000 light-years wide; its amorphous appearance certainly can be confused with a comet! "Not too difficult, but faint" I wrote for my first city-based observation

Could M60 have been the galaxy I saw from my bedroom window on the morning of February 25, 1968? I don't know. But during the predawn hours of that day, as I clambered along trying to set up my telescope in the frigid outdoor air, I realized that I was disturbing the rest of my family as they slept. There had to be an easier way on these really cold winter nights. "What about a room with a window to the east?" I wrote in my diary. "My own bedroom is in such a position, so I moved the desk from the window, and brought my 6-inch [later named Minerva] and its light weight mount from the basement, and placed it by the window.

"Looking through the telescope this way didn't quite work out. The tube's upper end still was affected by the heat from within the room and all the star images looked like comets. This was very dramatic and convenient but an inefficient way to observe. I then took away the mount and hand-held the tube so that its front end stuck out the window.

"I looked through the eyepiece. The stars were sharp points; all but one—a real fuzzy patch!

"I found in my *Skalnate Pleso* atlas that I was peering right into the heart of the Virgo galaxy group, and that one of the brighter ones probably is my 'new' object. But I can't be certain because I could not hold the scope steadily in one hand while I got out the atlas, all the way across the room, with the other. That would indeed be a tall order.

"Now, how to mount the scope. Perhaps, if I used a wooden block to hold the scope's rear end on the window sill, with a pulley system to pull the front end up and down it might work. Feasible but not very practical, and the stars would all swing about in the telescope field as if they were all singing in unison. I looked at the window sill; perhaps it had a suggestion.

"It did. Years ago my pencil sharpener was screwed into it, but now there were only two holes to prove it. Why not separate the mount of the telescope's lightweight stand from its pier, and attach it to the sill? I wouldn't even need to drill new holes.

"Ten minutes later this was completed. One hour had passed since I had given up on going outside, and now I was ready. From 2 to 3 AM I did a very pleasant hour of comet hunting."

I never did find out which galaxy brightened that morning, but it was either M60, one of the brightest in the Virgo group, or one of the other brilliant Virgo galaxies—perhaps M87?

Levy 175   NGC 4486 M87
Galaxy in Virgo
First seen: September 13, 1964
Position (2000.0): α 12 30.8 δ +12 24
Magnitude: 8.6
Distance: 57 million light-years
Best seen: in spring; observable in city sky
Classification: E0

Home to some five trillion stars, this galaxy is one of the largest in the Universe. M87 is about 120,000 light-years wide. Discovered by Messier in 1781, it impressed him as a trio of bright galaxies, with M84 and M86. What Messier couldn't know, of course, was that in its center was probably a black hole a *billion* times the mass of the Sun, with a strong jet of gases shooting out of it (which I actually saw one evening through the 61-inch telescope at Mt. Bigelow, Arizona). The Hubble Space Telescope image that cinched the black hole theory shows a spiral disk of hot gas in the galaxy's core. The disk is rotating rapidly, which is good evidence of a black hole in its center.

ELLIPTICAL GALAXIES

Far mightier than the spirals are the elliptical galaxies that can be overwhelming in size. Hubble classified the ellipticals from E0 (perfectly round) to E7 (very oval). Less than a fifth of the known galaxies are elliptical.

*Levy 176   NGC 4579 M58
Galaxy in Virgo
First seen: April 15, 1966
Position (2000.0): α 12 37.7   δ +11 49
Magnitude: 9.7
Distance: 53 million light-years
Best seen: in summer and fall; needs dark sky
Classification: SBb

> Once more the west was retreating, once again the orderly stars were dotting the eastern sky. There is certainly no rest for us on the earth. But there is happiness.
> —E. M. Forster, *Howards End*, 1910

Another massive elliptical galaxy, M58 is more than 90,000 light-years wide. Its huge number of stars does not prevent it from looking like a faint comet through a small telescope.

Levy 250   NGC 6946
Galaxy in Cepheus near Cygnus border
First seen: May 22, 2003
Position (2000.0): α 20 34.8   δ +60 09

Magnitude: 8.8
Distance: 23 million light-years
Best seen: in summer and fall; observable in dark sky
Classification: Scd

There are few experiences more fun for an amateur astronomer than spending a night at Kitt Peak. On the night of May 22, 2003, Wendee and I worked with Adam Block to photograph NGC 6946, an enchanting galaxy on the Cygnus-Cepheus border. Visually, it is a large, low surface brightness galaxy covering 11 by 10 arcminutes of sky. In photographs, the rich detail of its spiral structure begins to show, but Wendee and I had no idea of the intricacies that could emerge using a fine telescope, good guiding, and the photographic wonder of image processing. Once we pointed the telescope toward the galaxy, we exposed "dark frames" designed to capture and eliminate the "noise" that is inherent in any chip. We then took a series of ten-minute exposures of the sky, followed by single exposures in three colors. We finished the night by exposing "flat fields" to calibrate the exposures we had taken.

The result was an image worthy of Van Gogh, a magnificent "island universe." I couldn't believe my eyes—to get an image like that with a 20-inch, a small scope by Kitt Peak standards. How CCD technology has advanced! When I first started observing at Kitt Peak with the Planetary Science Institute twenty years ago, I doubt that an image like that would have been possible even with the 4-meter reflector. But regardless of the technology, that image would not have been possible without careful guiding and exquisite focusing. Given time for all those preparations, we barely had enough night to take this photograph.

As splendid as NGC 6946 is, its visual appearance leaves something to be desired, especially in a small telescope. It lies about a degree to the west of open cluster NGC 6939, a diffuse open cluster whose large size and greater integrated brightness (magnitude 7.8) might fool those with 3-inch telescopes into thinking that they have seen a remote galaxy when they have really just stumbled onto the cluster. But with more light-gathering power, more details of this wonderful galaxy begin to emerge.

Summer nights go by quickly, and before we knew it, the first sign of dawn tinged the northeastern sky. As Wendee and I prepared to walk back to the dormitory room assigned to us, we saw a bright satellite cross the brightening sky. I managed to turn my own 3.5-inch telescope on it, and even though it was racing through the field, I could clearly make out its oblong shape. Sure enough, we were watching the International Space Station pass by. I wondered if, in all the centuries that Native Americans have looked at the stars from this site, any could have imagined what happened here on May 22: we peered into the night to photo-

graph a distant galaxy, and then we saw another observing station, high above the earth, from which people were also watching the night.

Levy 338, PGC 48179, a galaxy 42 million light-years away. This figure is actually a composite of about seventy images of Comet Tempel 1 taken on July 3, 2005, the night that the Deep Impact spacecraft collided with it. The first one was taken at 09:54 PM MST, the last at three minutes past midnight. They show the comet as a trail moving from upper right to lower left. Notice how the trail brightens near the middle; this is when the comet's center went into outburst about twenty-five minutes after the impact, as if someone turned on a light in the inner coma. The brightness dims a bit at the left end because the comet is setting.

CHAPTER 12

FAINTER AND FARTHER
DISTANCES: SIXTY TO THREE HUNDRED MILLION LIGHT-YEARS AWAY

The very next day I woke up to see a sky blue enough to drink.

> —Robert Morgan,
> *The Muddy Chuckle*, 1964

We are like a Swedish navigator I knew once in Barcelona that had dreamed up a clever way of reckoning longitude by the stars and was uncommon accurate in all respects save one: to his dying day he could not remember whether Antares was in Scorpius and Arcturus in the Herdsman, or the reverse. The consequence of't was, he reckoned his lati-

tude by Antares with azimuths he'd sighted from Arcturus, and ran his ship into the Goodwin sands!
                    —John Barth, *The Sot-weed Factor*, 1960

As we soar farther and farther from home, we get to explore galaxies of all types and shapes, each one its own island in space. It is said that the total number of stars in all these galaxies is more than the number of grains of sand on all the beaches on Earth. As we move on through these galaxies, we are overwhelmed by the unending vastness of the universe.

Levy 19   NGC 6207
Galaxy in Canes Venatici
First seen: August 22, 1971
Position (2000.0): α 16 43.1   δ +36 50
Magnitude: 11.6
Distance: 56 million light-years

Levy 19, NGC 6207, lower left from the Great Cluster in Hercules.

Best seen: in summer; needs dark sky
Classification: Sc

If this galaxy were not so close to M13, the biggest globular cluster in the northern sky, it probably would get overlooked. But it's only 40 minutes to the northeast of the globular, which makes it easy to find. The star cluster is 23.5 *thousand* light-years away, and the galaxy 56 *million*. What does this mean? We are looking 23,500 years into the past, and 56 million years in the past, virtually at the same time. It is almost like looking out your Long Island window and seeing both your neighbor's house and Paris, France, instantaneously!

Walter Scott Houston (see chapter 4) told me about NGC 6207, that it was nicely juxtaposed next to Messier 13. The galaxy is so small, virtually indistinguishable from the starfield around it. And yet on the scale of the Universe, its 400 billion suns, clouds of gas and dust, and virtually infinite worlds make it a far more formidable object than the globular cluster. If the galaxy were as close as M13, it would fill a big part of the sky. Thoughts like that really make you look at the night sky in a profoundly different way.

I also enjoyed NGC 6207 on a clear, warm night while visiting Camp Minnowbrook on Lake Placid and again on the howling, stormy night of February 18, 2004. By observing in a small dome, I completely shut off the wind.

Levy 27   NGC 5377
Galaxy in Canes Venatici
First seen: August 6, 1970
Position (2000.0): α 13 56.3 δ +47 14
Magnitude: 11.2
Distance: 87 million light-years
Best seen: in summer and fall; needs dark sky
Classification: SBa

This barred spiral galaxy became famous briefly in February 1992 when a supernova appeared in it. The explosion probably took place only three days before it was detected by a telescope at the McDonald Observatory. SN 1992H, as it was called, the eighth supernova to be discovered in 1992, was a type II explosion that caused the star to brighten to 14 magnitude, almost as bright as the combined light of the rest of the stars in the galaxy *put together*!

Another kind of supernova, type 1a, is a "standard candle" for measuring the distances to distant galaxies. Since all type 1a supernovae have the same absolute blue magnitude of –19.6, they can be used in much the same way that the Cepheids are used. After a supernova flares, astronomers try to define which type it is. In the case of the supernova in NGC 5377, it was diagnosed as type II, which disqualified its being used for distance determination.

All this, of course, was far in the future when I first set eyes on this galaxy during my comet search at Camp Minnowbrook, on Lake Placid, on a clear summer night in 1970. It was an appropriate night to see a galaxy so far away. I had been discussing with another counselor whether the Universe had begun with a massive explosion or whether it was existing as a steady state, with matter being created at one end and destroyed at the other. Then I mentioned that five years earlier, Arno Penzias and Robert Wilson had detected the cosmic background radiation from that explosion, the "Big Bang." "I didn't know that," he said. "I think the Big Bang is right then." (More about that in the next chapter, but now back to supernovae.)

TYPE I SUPERNOVAE

The stellar explosions that can help with distance determination to far-off galaxies are type 1. In rare cases, it is possible for a star the mass of our Sun to become a supernova, but only if the star, in its white dwarf stage, happens to be one component of a double star. It may try to capture hydrogen from its neighbor. One consequence of this behavior can be a periodic nuclear explosion every few months such as we see in Tombaugh's star, or a blast every century such as with T Coronae Borealis (both stars are described in chapter 4). But what if the captured material never ignites? If the dwarf keeps on gathering more and more matter, how massive can it become and still have the stability that degenerate matter has in a white dwarf? (In degenerate matter, the nuclei of atoms are kept apart by a swarm of electrons that fight against the great force of gravity that is trying to collapse the star.)

Some decades ago, the Indian-born astrophysicist Subrahmanyan Chandrasekhar, who spent much of his career at Yerkes Observatory at Williams Bay, Wisconsin, proposed that there is a limit of mass beyond which a white dwarf cannot stay a white dwarf—the electrons in the degenerate matter can no longer keep the atomic nuclei apart—and after that point, it starts to contract again, becoming even denser. As soon as that limit is reached—1.4 times the mass of the Sun—the star blows up. If the conditions are right, a white dwarf can end its life spectacularly after living its normal life as a star for many billions of years. It is even possible for a single white dwarf star to become a supernova, if its mass exceeds Chandrasekhar's limit.

TYPE II SUPERNOVAE

There are apparently at least two versions of the type II, depending on how massive the star is. In either case type II explosions are the end result of stars that have lived too hard and too fast. Burning themselves out in just a few million years, compared to the ten-billion-year lifespan of a star like our Sun, these stars end their lives with great violence. In the weaker type, the star fuses its hydrogen, and then its helium, until the core is left with carbon. Like the helium flash in sunlike stars, there is a carbon detonation when all the carbon in the core ignites at once. This detonation may be strong enough to blow apart the supernova's core.

More massive stars survive carbon detonation, but in doing so, they have bought only a few hundred years of time. Stars more than nine times the mass of the Sun are so hot that their carbon ignites gradually. A very massive star twenty-five times the Sun's mass, after spending seven million years fusing hydrogen to helium, alters its helium supply into carbon in half a million years. The process of carbon ignition begins gradually, so it is safely fused to oxygen in about six hundred years. Then the oxygen becomes silicon in the short time of about six months. At the end, in less than a day, the silicon fuses to form a core of iron.

If stars only understood nuclear physics, they'd know that an iron atom is so stable that it cannot undergo the process of nuclear fusion. Instead of releasing energy in a fusion reaction when enormous amounts of heat are applied, iron will absorb the energy. But obviously a star doesn't understand this, so it tries to ignite its iron core anyway by contracting and heating it. Unable to keep up, the core suffers a final collapse. In less than two seconds, it crashes in on itself, carrying large amounts of still-unused fuel; as the electrons crash into the nuclei of their atoms, they form neutrons and neutrinos, and a new kind of star called a neutron star.

Around the collapsing stellar core, bedlam reigns. A shock wave pushes the star's outer layers away at tremendous speed. In the titanic explosion, the star might outshine the combined light of all the stars in its galaxy. This is the explosion that took place 87 million years ago and became visible from the Earth's sky in 1992.

*Levy 37   NGC 3055
"Sleeping Cat" Galaxy in Sextans
First seen: 1981
Position (2000.0): α 09 55.3   δ +04 16
Magnitude: 12.1
Distance: 83 million light-years
Best seen: in winter and spring; best in dark sky
Classification: Sc

Levy 37, Sleeping Cat Galaxy in Sextans.

Looking rather more like a small comet than a distant star-filled galaxy, NGC 3055 has a split personality when seen in photographs. On the east side, its arms spread out evenly like any other self-respecting galaxy we see from the respectable distance of 83 million light-years. But its west side—the cat's head—is a confused jumble of shorter arms and spurs from arms. Some past interaction with another galaxy, which didn't collide but passed a bit too close, could have disrupted this galaxy.

Whatever happened and continues to happen, NGC 3055 is worth a look. Don't be afraid to insert a higher-power eyepiece to see the magic of this spiral.

Levy 106   NGC 4038 and NGC 4039
Colliding galaxies in Corvus
First seen: 2001
Position (2000.0): α 12 01.9   δ −19 52
Magnitude: 10.7
Distance: 66 million light-years
Best seen: in spring; needs dark sky
Classification: Spiral irregular

> A pair of star-cross'd lovers take their life . . .
> —William Shakespeare, *Romeo and Juliet* 1.1.6, 1595

Levy 106, NGC 4038–4039. A view of what the Milky Way and Andromeda Galaxy might look like when they collide, far in the future.

It's a fate that begins with a "pass-through" collision in which two galaxies slide through each other. Being so far apart, the stars survive, but huge clouds of dust trigger vast episodes of star formation in both galaxies. As the galaxies move apart, their mutual gravitation refuses to let them get too far away from each other. They swing around in a sort of macabre cosmic square dance and circle toward each other for a second encounter. This time, long tendrils of stars escape into space as the two galaxies begin to merge. When it's over, a single new phoenix of a galaxy, an elliptical, is born of the rubble.

Haven't we heard this story before? Isn't this the predicted collision between the Milky Way and Andromeda Galaxy, already mentioned in chapter 10? Yes, it could be. But this time I tell a story of what happened "long, long ago," like in *Star Wars*, in two galaxies far, far away. The far-off galaxies are NGC s 4038 and 4039, and we are seeing a collision that took place 66 million years go, about the same time that a comet or an asteroid collided with Earth. It just took their light that long to get here and tell us about it.

Variously titled the Ringtail, Rattail, or the Antennae, here we see an event more than an object, a study in two galaxies merging. Even in small telescopes (like a 6-inch reflector), the galaxies are visible as a single complex bar with NGC 4038 on the northwest and NGC 4039 on the southeast. With your telescope, I suggest that you use a medium-power eyepiece (16 mm to 12 mm) to enjoy the details of this extraordinary and distant joining of two islands of stars and dust.

Levy 128P   IC5020
Galaxy in Microscopium
First seen: October 7, 2002
Position (2000.0): α 20 30.6   δ −33 29
Magnitude: 13.0
Distance: 135 million light-years
Best seen: in fall; best in dark sky
Classification: Sa

I added this galaxy to our catalog not for what it is but for where it is—part of a line of foreground stars, part of which looks like a question mark. It was one of the more joyous finds in the Shoemaker-Levy double cometograph phase of our comet search.

THE SHOEMAKER-LEVY
DOUBLE COMETOGRAPH

After we ended our Palomar Asteroid and Comet Survey in 1994, Carolyn and Gene Shoemaker and Wendee and I set up a new program using a series of small Schmidt cameras. The photographic mode of the program consisted of a set of twin 20-cm F/1.5 Schmidt cameras, manufactured by Celestron International but improved and refitted by Epoch Instruments. Each camera was fitted with a special Vehrenberg film holder allowing a coverage of ten degrees of sky in a circular field. Although the two cameras shared a single mount, Ophelia 2 saw a field of sky centered about ten degrees north of what Ophelia 1 peered at.

We used an old atlas that Dad gave me in 1964, with 10-degree fields (100 square degrees each) marked off. Fields 1–175 are the nova search fields straddling the Milky Way and created by the American Association of Variable Star Observers; fields 176–428, covering the rest of the sky, are comet search fields created by Jim Low of the Royal Astronomical Society of Canada.

The most important part of comet searching is to develop a strategy. With film, the plan should be designed to cover as much area of sky as possible without covering areas that are already well searched by the professional surveys. Our program searched the sky within 90 degrees of the Sun, and south of –30 degrees, which allowed us to pick up galaxies like IC5020.

Here's how we did it: After Wendee carefully prepared a 2.25 by 2.25-inch film, writing its number on the edge, in the dark, and loading it into the film holder, I would load it into one of the telescopes and center the star on which I would guide the next eight-minute exposure. Once the guide star was centered, Wendee *gently* removed the cover from the telescope, then held it a short distance from the camera while Carolyn or

I made final adjustments on the position of the guide star. When all was ready, Wendee moved the cover out of the way, and the exposure began. Wendee recorded the time to the nearest second, and I read to her the telescope's current position, along with a description of the stars in the field, so that we could return to the same guide star when it was time to repeat the exposure. On a typical evening, we photographed four or five pairs of photographs using both Ophelia cameras, for a total of sixteen or twenty pictures.

After we took an exposure, we then photographed a second field, then a third, a fourth, and sometimes a fifth. When the series was over, we then repeated all the exposures. Ideally, between forty-five minutes and an hour separated each pair of exposures.

The night's exposures done, they were photofinished using a high-contrast developer like Kodak's D-19, then fixed, finally washed for thirty minutes, and allowed to dry. Carolyn or I then scanned the films with an instrument called a stereomicroscope in which each eye would concentrate on a different film. Since the films are identical except for the time, they appear the same. Only a moving comet or asteroid would give itself away by appearing to "float" above the background of stars.

OF FOCUS, FRUSTRATION, AND MONGOLIAN DUST

It took us over a year to focus Ophelia 2, one of the most aggravating and difficult, yet fun, experiences I've ever had in astronomy. Ophelia uses special rods, which do not change their lengths with temperature, to support the film holder and to keep it at a fixed distance from the primary mirror. So once the telescope is focused, it should stay focused forever.

A whiz of a telescope optician, our friend Bob Goff succeeded in focusing one of the cameras, Ophelia 1, in his lab without difficulty. But when Ophelia 2 stubbornly resisted, he decided to focus her under the stars. We began one clear evening by taking, developing, and examining a single photograph. Bob used a wrench to adjust each of the three pairs of locking bolts the tiniest fraction of a turn, and we repeated the exposure. We improved the focus after a few tries and continued the project some nights later.

Focusing Ophelia 2 almost became an end unto itself as we continued the slow and painstaking process. It also had some unintended consequences. One evening, while hurrying to meet Bob, I got a speeding ticket. To avoid a fine, I had to take an all-day course in driving safety. Wendee, who audited the course with me, noted that the instructor had a hard time believing that the reason I got the ticket was that I was rushing to get home by nightfall to start focusing a telescope.

On a happier night, while facing the telescope pointed to the zenith and standing atop a ladder with wrench in hand, the world-renowned telescopist started answering my questions using a perfect imitation of Donald Duck. "How close are we to focus?" I asked. "*Close—er!*" Bob quacked. Those nights, sadly, had to come to an end when Bob passed away. By this time we easily could have abandoned Ophelia 2 altogether—after all we did have two well-focused instruments. But we wanted to finish the job partly out of respect for Bob, whom we really miss.

Not long after Bob's memorial service, Dean Koenig, another good friend and telescope expert, resumed the focus effort. This time Dean and I worked all night long, taking one film after another. We tried to use the Orion Nebula, M42 (see chapter 6), for our test photos since the nebula was Bob's favorite object. To save time, I moved the photofinishing chemicals out into the observatory, and we kept the developing and fixing times to the absolute minimum needed to show images. I also took off my wedding ring—a Gibeon meteorite—

so chemicals wouldn't get all over it. I put it down somewhere in the observatory, and, years later, I still have not found it. With eight-minute exposures, ten minutes' developing time, five-minutes of fixing, plus time for examining the films, and more time to adjust the focus, we were lucky to get three exposures of M42 per hour.

We repeated this process over five or six nights spread out over several months. On one night, the sky began brilliantly clear but deteriorated as the hours went on, as if huge amounts of dust were being kicked up into it. But the air was calm. It wasn't until the next morning that we heard that a windstorm from Mongolia—*Mongolia!*—had sent a large cloud of dust into the atmosphere, across the Pacific, and inland to Arizona just in time for our focus session.

On another night one side of the film was sharp while images on the opposite side were elongated. Dean finally got the elongations to a minimum. We could have stopped there, but we decided to see if we could improve the focus further. The next adjustment turned out to be in the wrong direction; when we turned the nuts backward, another side began to go out, and by the end of the night our films sported doughnut-shaped star images, the signature of a seriously out-of-focus telescope. We stopped, although an ever-optimistic Dean kept saying, "Don't worry, we'll get it yet! We got a lot of information tonight about how the telescope behaves. We'll get it." Dean was right of course; with each turn of the nuts supporting one of the telescope's three invar rods, and seeing the result, we were knowing our telescope better and better. We ended the night with a collection of many examples of poorly focused and misaligned star images.

At last, the great night came: April 25, Session *12820EM2. Carefully making notes of every step he took, Dean made change after incremental change. Each one brought the telescope closer to focus. Late in the night, while dangerously close to focus, we had to make a fateful decision. Should we continue trying, and risk losing what we had accomplished? We decided to continue, and this time the effort happily paid off. We ended the night with a telescope focused better than it had ever been.

What lessons did we learn from this three-year-long exercise? The most important one is that you must get to know a telescope, its behavior, and its habits. Our case was an extreme one—it took many hours to adjust the scope so that it would produce consistently good results. But somehow, when we examine a picture taken by this camera, we appreciate it much more. It is a source of continual amazement to me that I can focus our CCD camera automatically, in less than three minutes, whenever I choose to. But more about our CCD search later in this chapter.

*Levy 180   NGC 4303 M61
First seen: July 16, 1966
Position (2000.0): α 12 21.9   δ +04 28
Magnitude: 9.7
Distance: 62 million light-years
Best seen: in spring; needs dark sky
Classification: Sbc

Levy 180, M61.
Tim Hunter photo.

THE GREAT IDEA OF A MESSIER CLUB

When I first arrived at the Royal Astronomical Society of Canada's Montreal Centre on October 8, 1960, I quickly learned that Isabel Williamson, one of its most prominent members and leader of its observation program, had founded North America's first Messier Club in the early 1940s. The idea of the club was a friendly competition to see the Messier objects. "Its main purpose," she wrote, "was to stimulate members into becoming active observers instead of being content to look through the telescope at objects that others had located." The club had rules: to receive credit for sighting a Messier object, the observer would not look through someone else's already pointed telescope, but would have had to find it, preferably by star hopping using the telescope and finder. In the earlier years, setting circles were prohibited, but that rule was relaxed later. (I wonder, though, if Miss Williamson would have thought that using a computerized telescope with "go to" capability would have counted.) So as I recorded object after object, I sent my notes to her, and my sighting would be duly recorded in the center's Messier notebook and on the large board that showed everyone's totals. I was the tenth graduate of the Montreal Centre's Messier Club.

I had trouble with M61, though. A large, faint galaxy in Virgo, I found it the most difficult of the Messiers to spot. It is hard to see because it is a "low surface brightness" galaxy; that is, it might be brighter than 10 magnitude, but its light is spread out over a large area (6 arcminutes), and it has no appreciable brightness increase toward its center—at least it doesn't at low power. Thus, it is hard to see unless the surrounding sky is very dark. "Very very difficult to see for first time," I wrote. "Even though conditions were excellent, I found M61 extremely difficult to make out. I think it is the hardest of the Messiers I have seen so far."

Here is where the idea of a competitive Messier club makes so much sense. Miss Williamson compared my observation of M61 with those of other members who had reported it earlier. "Upon referring to the Messier Club notebook," Miss Williamson wrote in the center newsletter *Skyward*, "we gather that other observers didn't find M61 very easy either. . . . Klaus Brasch, using an 8-inch reflector, reported it as 'very faint and diffuse.' . . . Our records also show that Geoffrey Gaherty observed it with his 8-inch reflector," but that all the observers found it difficult.[1]

The modern Montreal Centre Messier Club would do well to add this note from Steve O'Meara, who racked up his eyepiece power to 130 to see a nucleus that resembles a bright star surrounded by a quadrilateral of light surrounded by knotlike regions where stars are being formed: "Overall, at high power, the galaxy looks like a square with slightly rounded edges," he wrote in *The Messier Objects*. And he elaborates, "With averted vision, the northernmost knot shows an extension that trails off irregularly to the east in a wavelike fashion, like a gracefully thin wave of smoke. Closer to the nucleus and to the south is another star-studded region that trickles off on lumps to the west before it curves sharply to the north. A faint but definite arm can be glimpsed to the east, and it boxes in the nuclear region." Not for nothing did Steve earn the title CCD-eyes! It is a testament to his powers of observation and patience that he is able to gain so much out of a distant galaxy that appears so featureless to less observant eyes.

Levy 210   NGC 4685
Galaxy in Coma Berenices
First seen: November 13, 2002
Position (2000.0): α 12 47.1 δ +19 28
Magnitude: 12.6
Distance: 293 million light-years
Best seen: in spring; needs dark sky
Classification: S0
"Winking Galaxy"

> Stars, hide your fires,
> Let not light see my black and deep desires.
> —Shakespeare, *Macbeth* 1.4.49–50, 1606

Just like some of the planetary nebulae I have described (see chapter 7—NGC s 2392 and 6826), this galaxy winks; if you concentrate on its bright core, the rest of the galaxy disappears. To view this effect most clearly, I used a medium-power, 16 mm eyepiece. It's one thing to observe this effect with single stars within our own galaxy, but to see it involving a whole galaxy with a bright core of light at its center is another experience altogether.

Levy 319P   ESO573-12
Galaxy in Corvus
Added: December 17, 2004
Position (2000.0): α 12 20.6   δ −18 40
Magnitude: 14.0
Distance: 350 million light-years
Best seen: in spring; needs dark sky
Classification: Sbc

Levy 319P, ESO573-12, galaxy at left. Levy's Devonian turtle fossil was alive and crawling about when light from the galaxy left on its journey to Earth.

... Come, gentle night, come, loving, black-brow'd night,
Give me my Romeo, and, when he shall die,
Take him and cut him out in little stars,
And he will make the face of heaven so fine
That all the world will be in love with night,
And pay no worship to the garish sun.
—Shakespeare, *Romeo and Juliet* 3.2.20–25, 1595

I added this galaxy for a simple reason; it is located a short distance from Clyde Tombaugh's star, my favorite variable star. It's a case of pure familiarity: on night after night, as I have photographed the variations of a relatively nearby star, I have encountered a spiral galaxy, not unlike our own Milky Way but 350 million light-years from us. Around what star with membership in that distant orb might there be a planet like Earth?

On September 27, 1969, I joined two friends, Mikael Stoffregen and David Roy, on a geological field expedition to the Gaspé shore of northern New Brunswick. On that raw, damp day, we mapped the stratigraphy of a Devonian rock structure called the Escuminac formation. We also collected some sample sandstone. One, about the size of a thick pancake, had some fossil structure in it.

Back at our lab at Acadia University's geology department, David cut the rock open. When he pulled the two pieces apart, an almost perfect specimen of fossilized turtle (see figure on page 235) revealed itself to the light of day for the first time since it died some 360 million years ago. That small prehistoric animal crawled about at the time that the light from ESO573-12 began its long journey to Earth, long before *Brontosaurus* and *Tyrannosaurus Rex* roamed our planet.

On that world, in that galaxy so far away, would anyone be gazing back at a tiny, 14 magnitude blob of fuzzy light, wondering if there were an "us" here? Or would life there be limited to small turtles sticking their little necks out and peering up at the sky?

Levy 327P   IC2531
Galaxy in Antlia
First seen: October 7, 2002
Position (2000.0): α 09 59.9   δ −29 37
Magnitude: 12.5
Distance: 100 million light-years
Best seen: in winter and spring; needs dark sky
Classification: Sb

Before he moved to California in later life to discover objects like IC2531, Louis Swift had a most interesting career as a comet hunter. He discovered the first of several comets on July 15, 1862, using a 4-inch refractor telescope from his home in Marathon, New York, in the midst of an educational region; six years after the discovery, what is now the State University of New York at Cortland was founded only sixteen miles to the north. Though the comet was independently discovered by another American amateur astronomer, Horace Tuttle, while in England, it was called Rosa's comet, for the observer who first saw it from Rome on July 25.[2] Although the comet itself returned in 1992, we can see its effects every August as the great Perseid meteor shower.

IC2531 is an example of a typical Sb spiral. In later years, Swift relocated to California to observe at the Mount Lowe Observatory at Echo Mountain, near Los Angeles, where he spent seven years, from 1893 to 1900. IC2531 is the eleventh galaxy he listed while there.

This galaxy has recently been the subject of a study to determine how much dust Milky Way–type spirals contain and where that dust lies. IC2531, observed in infrared light, helped lead to some general ideas about the opacity of these galaxies. The most interesting result is that spirals have less dust than was previously believed; if they were observed face-on (like M33 and M51), we could see right through them. What dust they have is concentrated, not surprisingly, at the plane of the galaxy's disk.[3]

Levy 331P   NGC 3319
Galaxy in Ursa Major
First seen: January 17, 2005
Position (2000.0): α 10 39.2   δ +41 41
Magnitude: 11.3?
Distance: 61 million light-years
Best seen: in winter and spring; needs dark sky
Classification: Sc

Levy 331P, NGC 3319.

A galaxy hung over from staying up too late? When I saw it on four images taken on January 17, 2005, the east side of NGC 339's spiral arm structure appeared graceful, almost in flight. The west side is filled with what look like H II (H two) regions, distant stellar nurseries. Moreover, the western arms seem to be pulled or stretched out into space more than the eastern ones. The center is strongly elliptical. In any case, this is one magnificent galaxy to photograph.

THE JARNAC OBSERVATORY
COMET SURVEY

In the description for IC5020, I wrote about the photographic program Carolyn Shoemaker, Wendee, and I conducted, starting in 1995. With NGC 3319 I get to share what that phase of our program evolved into. Compared to the rigor and effort in a photographic comet search, an automated hunt using CCDs seems positively joyful in its ease. Well, not really; it is still a lot of work, only the work is different. No longer must we stay outside, braving the elements. Most of the work can now be accomplished from a climate-controlled room. Happily, no more need to be outdoors in cold weather! Sadly, no more need to be outdoors and under the stars.

Today we now can create a list of fields to photograph, and the telescope does the rest without the need for any supervision except to open and close the observatory and the telescope.

Combined with automated scanning of the images, it means that comet hunting can be done in your sleep, somewhat as in the *San Francisco Examiner* article, from March 8, 1891, which suggests that the great comet hunter Barnard could find comets in his sleep. In that article, which had been sent to the newspaper as a hoax, a special device would examine the spectrum of everything as Barnard slept. "Stars, nebulae and clusters innumerable crowd into the field [of view of the telescope] with every advance of the clock," the article went on, "but the telescope gives no sign of their presence." But should any object showing the "three bright hydrocarbon bands" of comet light appear, the device would allow that light to pass through and hit a selenium band, closing a circuit and setting off an alarm in Barnard's bedroom. The sleepy astronomer would rush upstairs to the telescope, make a simple visual confirmation of the new comet, report it, and return to bed.[4]

That old nineteenth-century hoax isn't quite how we really search now, but the idea is similar. Our telescope doesn't take automated spectra; instead it looks for moving objects on multiple images. Before we could boast of some easy searching, Wendee and I, with a lot of help from programmers and telescope experts, spent almost a year building the system and working out its bugs. Connecting telescope to camera to computer was a difficult process that finally started to work after a lot of trial and error.

One way of searching is called "fences." It works when the CCD chips do not have very large fields. We let the sky itself widen the search area. We cast a net across a portion of sky. Each night we take a strip of fifteen photographs for a total area coverage of 3/4 by 7 degrees. The same area is photographed three times, and we repeat the entire procedure on following nights. Assuming that a typical comet moves half a degree per day over a month, we are essentially covering an area of 15 by 7 degrees. By doing two such areas per night, and using two telescopes, we increase the coverage by a factor of four, to 60 by 7 degrees. Moreover, since the program mostly runs itself, it has become a part of our daily routine.

As we put more telescopes and wider-field CCD cameras to work, we can increase our area coverage. With CCD cameras now allowing fields of 1 degree or more, we can search for comets in much the same way we used to work photographically. NGC 3319 was found while Clyde, our 14-inch telescope, was busy photographing area 377. With film, we photographed the entire area in a single eight-minute exposure, repeated once. With CCD, we take about ninety exposures to cover the same area—a much longer time, but with a bigger reward—the search, now automated, takes us to objects one hundred times fainter.

CHAPTER 13

OF SPACE, TIME, AND DREAMS
DISTANCES: HUNDREDS OF MILLIONS TO SEVEN BILLION LIGHT-YEARS AWAY

These friends, like astronomical distances,
are only to be spoken of in the very largest
figures.

—Charles Dickens,
Our Mutual Friend, 1865

As we travel out in space, and further back in
time, past galaxies and clusters of galaxies, to
distant quasars, is it possible that our journey
can take us back to the beginning of time itself,
the explosion that started the Universe? What
would it look like? We are now so far away from
home that we see clusters of galaxies racing

around other clusters of galaxies, the first evidence of the expanding Universe.

Following the tradition of earlier chapters, this chapter considers first the distant clusters of galaxies in order of their inclusion in the catalog and then the more distant and enigmatic quasars.

Levy 51   Arp 321
"Larry, Moe, and Curly" galaxies in Hydra
First seen: 1982
Position (2000.0): α 09 38.9   δ −04 52
Magnitude: 12.0
Distance: 281 million light-years
Best seen: in spring; needs dark sky
Classification: Two are lenticular, the brightest is elliptical, and the two
    faintest are spiral Sb.

No matter how good your observing records are, they can always be better. Years ago, while hunting for comets, I came across what looked like a single three-sided galaxy. Such a sight needed to be explored with a higher power, and so when I inspected the object with a 13 mm Nagler eyepiece, it split into three distinct galaxies. So odd did this trio appear that I dubbed it Larry, Moe, and Curly after the Three Stooges. But somehow I never bothered to write down their position in Hydra or mark them on an atlas.

I loved the subtle splendor of this trio, and I began to show them off at star parties and other observing sessions. On the Saturday evening of September 1, 1984, I recorded in Session 6701AN that Gerry Rattley, Gene Lucas, and other amateur astronomers had made the two-hour drive from Phoenix to observe with me. In my log I noted being "especially impressed with Gerry Rattley's ability to observe faint objects." These are people who know the sky so well that its details are imprinted in their minds. I told the group about the Hydra trio. "I

Levy 51, Arp 321, "Larry, Moe, and Curly" and two friends in Hydra. Photograph by Tom Glinos using a 25-inch Ritchey-Chrétien telescope and Finger Lakes "Dream Machine" CCD.

remember how quickly you were able to pick it up in the 16-inch," recalled Gerry of that night. The rest of the group looked at the weird-looking faint fuzzy at low power then marveled at the three-galaxy split at high power. When I wrote up that session, I carefully noted that we had observed "Larry, Moe, and Curly in Hydra," but inexplicably I again failed to note the trio's position. Over the years I went on to other things, and like an old friend out of touch but not forgotten, I stopped looking at it.

In the spring of 2000, I was getting my thoughts together on my particularly favorite objects, especially ones found during comet hunting, as a prelude to this book. Larry, Moe, and Curly seemed a natural entry. On returning home I began a search, first through my special catalog of comet masqueraders. When I didn't find them there, I looked through volume after volume of observing logs. Though I came across Larry, Moe, and Curly several times, not once did I record their position.

In the early months of 2001, with Hydra high in the south, I began a search through the 16-inch, looking where I thought it was. After two fruitless hours I gave up. By now I was really puzzled. Could I have made the whole thing up? How could I have so completely forgotten the location of one of my favorite objects? I thought the object was lost forever among the hundreds of faint galaxies in Hydra.

Six months later, after a lecture I gave to the Phoenix Astronomical Society, Gene Lucas turned up in the audience. It was good to see him again after all these years and to learn that both he and Gerry were still avid observers. Naturally I asked Gene about the Hydra trio, and the next day he sent me the e-mail I was waiting for. Gerry, it turned out, knew where Larry, Moe, and Curly were.

Back in 1984, when the group awoke around noon after our all-nighter, they unexpectedly received an invitation to spend an evening with Bob Goff (the man who began the focus work on Ophelia described in chapter 12) at Kitt Peak. Not wanting to miss an opportunity to observe at one of astronomy's holiest sites, Gerry didn't turn him down. While up there, he visited the library, looked up the position he had recorded for Larry, Moe, and Curly, and found at that spot no. 321 in Halton Arp's catalog of interesting galaxies. It is actually a cluster of five galaxies, not three, though two fainter elongated spirals are below my limiting visual magnitude. Unlike me, Gerry carefully recorded the information, and seventeen years later he still had it.

A lot of life has passed by between that long ago session and No. 12458M3, on the morning of October 15, 2001. Just hours before the birth of our grandson Matthew, I set up Miranda to begin session 12458M3. Arp 321—aka Larry, Moe, and Curly— appeared in the eyepiece just as it did so long ago. An old celestial friendship was alive again, added to my list of objects spotted during comet search. This time, I carefully noted its position in my observing log, my lesson learned at last.[1]

Levy 158   "Castor Cluster"
Galaxy cluster in Gemini
First seen: 1985
IC2193 Position (2000.0): α 07 33.3 δ +31 27
IC2194 Position (2000.0): α 07 33.7 δ +31 19
IC2196 Position (2000.0): α 07 34.1 δ +31 24
IC2197 Position (2000.0): α 07 34.3 δ +31 20
IC2199 Position (2000.0): α 07 34.9 δ +31 13
Magnitudes: about 13
Distances: IC2194 is 187 million light-years; IC2197 is 204 million light-years; IC2199 is 202 million light-years
Best seen: in winter; needs dark sky
Classifications: IC2194 is Sc; IC2196 is E0

> The historiographers repeate that Castor and Pollux have been often seene in battailes sitting on white boxes, valiantly fighting against enemies campe. Beyond the mythological reference of the battle of Lake Regillus, the twins are also the brightest stars in Gemini, a prominent winter constellation.
> —Lewes Lauaterus,
> *Of Ghosts and Spirits Walking by Nyght*, 1572,
> in a mythological reference to the battle of Lake Regillus

How can there be a group of galaxies so close to the Milky Way? I asked myself in the early months of 1985. My first response: there couldn't be. I was pretty familiar with this particular region of sky, between Pollux and Castor, the twins of Gemini. It provided my first five minutes of comet hunting on December 17, 1965, and in the years since then, I have often returned to that region. At no other time did I ever spot any faint fuzzies there.

The second thought, the possibility that I had discovered a cluster of comets, certainly came to mind. It had happened before. The great sungrazing comet of September 1882 had an effect, both dreamed and real, on the great observer E. E. Barnard. Early in the morning of October 14, he dreamt about comets appearing all over the sky. Barnard awoke in the last hour before dawn, went outdoors, and set his telescope on the rising comet. He studied it briefly before beginning a scan for new comets. Moving his telescope to the southwest, he didn't have far to go—perhaps five or six degrees—before he found a group of a half dozen small comets. Flabbergasted, Barnard wondered whether he had fallen asleep on his feet and had resumed his dream. The comets were real—the observation was confirmed the following night by observers in Europe. But although the little comets were traveling at the same rate and direction as the main body, they all disappeared in less than a day.

What was this grouping near Castor? After an hour of observing, these objects were still plastered in the same positions in the sky, so they didn't appear to be comets. Finally, I identified them as faint galaxies on a star atlas and informally called them the "Castor Cluster."

Levy 200 Fornax cluster
Cluster of several galaxies in same field
First seen: 1982
NGC 1380 Position (2000.0): α 03 36.5   δ −34 59
NGC 1399 Position (2000.0): α 03 38.5   δ −35 27
NGC 1404 Position (2000.0): α 03 38.9   δ −35 35

Magnitudes: 10.0–11.0

Distances: NGC 1380 is 67 million light-years; NGC 1399 is 63 million light-years; NGC 1404 is 67 million light-years

Best seen: in fall; needs dark sky

Classifications: NGC 1380 is lenticular; NGC 1399 is E0; NGC 1404 is E0

ORIGINS

As we go back in space these hundreds of millions of years, we also go back in time. The light of the Fornax cluster left when dinosaurs roamed the Earth. With telescopes like the Hubble and projects like its Deep Fields, we can see galaxies several times farther away. We can even contemplate the beginning of the Universe. If the Big Bang theory (so named derisively by the British astrophysicist Fred Hoyle, but accepted by most astronomers) is correct, the beginning might have looked like this:

A sea of quarks, antiquarks, and photons occupied the Universe just before its first thousandth of a second. It was a sort of primordial soup in which quarks combined into pairs and triplets, and then into protons and neutrons. Quarks and antiquarks—matter and antimatter—both existed in almost equal numbers at this time. The "war" between them was won by a small surplus of quarks by the end of the first second, and the quarks began to gather to form protons and neutrons. The electromagnetic force came into play at this time. This force is much stronger than gravity. However, because protons and electrons have precisely equal and opposite charges, the electromagnetic force normally cancels itself out. If the charges were not equal, the electromagnetic force, and not gravity, would have determined the large-scale structure of the Universe. There would have been no galaxies, no stars, and no planets, just unformed energy and lightning.

That violent things more quickly find a term
Is shown through Nature's whole analogies;
And how shall the most fierce of all be firm?
    Would you have endless lightning in the skies?
                    —Lord Byron, *Don Juan* 14.94, 1823

We now scan forward just a few minutes. Although the Universe was cooling rapidly, it was still as hot all over as the center of a star. During these minutes, protons and neutrons fused together to form "heavy hydrogen" (deuterium) and alpha particles, or helium nuclei. But with a continuation of the cooling process, within a few minutes at most, fusion shut down.

Now we fast-forward about three hundred thousand years. From the end of the first three minutes to the end of the Universe's first three hundred thousand years of existence, there was but a single item on the Universe's agenda: cooling down. By that time the temperature of the entire Universe had dropped to less than 3,000° F, and it became transparent to photons of electromagnetic energy. They were set free, and for the first time there was light. We can now detect these primeval photons as something called background radiation. It is the earliest indication we have of the early universe, for it represents the time that it first became detectable.

We now race ahead 750 million years, to a galaxy far beyond the Abell 2218 cluster of galaxies. This galaxy is visible to us only because the cluster, originally cataloged by George Abell, acts as a gravitational lens, magnifying its distant light. Announced in 2004 by a team led by Jean-Paul Kneib and others using the Hubble Space Telescope and the twin Keck telescopes atop Mauna Kea, Hawaii, this galaxy, more than others, might be partially a relic of the past—probably none of its stars still exist.

Levy 219 Hydra I Cluster
First seen: November 17, 2002

NCG 3309 Position (2000.0): α 10 36.6   δ –27 31

NGC 3311 Position (2000.0): α 10 36.7   δ –27 32

NGC 3312 Position (2000.0): α 10 37.0   δ –27 34

NGC 3314 Position (2000.0): α 10 37.4   δ –27 41 (actually two spirals!)

NGC 3316 Position (2000.0): α 10 37.6   δ –27 36

Magnitude: 14.0

Distances: NG0C 3309 is 158 million light-years; NGC 3311 is 157 million light-years; NGC 3312 is 155 million light-years; NGC 3314A is 156 million light-years; B is 160 million light-years; NGC 3316 is 157 million light-years

Best seen: in spring; needs dark sky

Classifications: NGC 3309 is E0; NGC 3311 is lenticular; NGC 3312 is Sb; NGC 3314A is Sab; NGC 3314B is Sc; NGC 3316 is lenticular

[T]he distance of the starrie skie is from us, in Semidiameters of the Earth 20081½. Twenty thousand fourscore, one, and almost a half.

—John Dee, *Euclid's Elements of Geometry*, circa 1600

Also called Abell 1060, the Hydra 1 cluster contains some one hundred galaxies spanning about three degrees of sky. NGC 3314 was considered a peculiar spiral until it was imaged by the Hubble Space Telescope in 1999 and 2000. The images showed two normal spirals, one 4 million light-years directly behind the other, merged together on our line of sight. The chance alignment gives force to the new idea that spiral galaxies might have sufficient dust to hide as many as half their stars.

With a dark sky, these galaxies can be found fairly easily in an 8-inch telescope. NGC 3309 is a magnitude 12.5 spiral, slightly elongated with a bright core. NGC 3311, at magnitude 11.6, is brighter, but lacks the bright core. NGC 3312 appears larger than the others, covering 3.4 arcminutes in diameter. NGC 3316 is quite a bit smaller.

The Hydra cluster is probably part of the Hydra-Centaurus

supercluster of galaxies. Beyond that, there is something entirely new called . . .

THE GREAT ATTRACTOR

We know that all the superclusters in the Universe are racing away from one another. But one interesting discovery shows that the galaxies in our neighborhood, including the entire Local Supercluster, seem to be heading toward a distant supercluster, or pair of superclusters, in Centaurus. Composed of many thousands of galaxies, the Attractor is about 400 million light-years wide and is spread out over a large area in the southern sky. It is at least 100 million light-years farther away than Hydra 1. It contains the mass of more than 10 quadrillion (10,000 trillion) suns, which is thirty times more massive than the Local Supercluster. It defies visibility because its light is blocked by the dust of our own galaxy—the southern regions of the Milky Way.

"The Seven Samurai" are responsible for the discovery of the Great Attractor—David Burstein, Roger Davies, Alan Dressler, Sandra Feber, Donald Lynden-Bell, Roberto Terlevich, and Gary Wegner. Their work confirmed an assumption proposed by Clyde Tombaugh as part of his great sky survey from 1929 to 1945, that galaxies and clusters of galaxies are not evenly distributed.

THE GREAT WALL

In 1989 astronomers Margaret Geller and John Huchra of the Harvard-Smithsonian Center for Astrophysics reported a "wall" of galaxies between 250 million and 500 million light-years across! The Wall stretches through the Northern Hemisphere spring sky. Their method was to chart the distribution of galaxies in space in three dimensions, not the two that viewers of the sky normally use. Using this technique, Geller and Huchra have found that the superclusters lie on what looks like sheets atop bubbles. The largest "bubble" so far found is the Great Wall. In a different survey, the University of Hawaii's Brent Tully discovered that roughly one hundred large and rich clusters of galaxies lie in a disk-shaped structure that is 1.5 billion light-years long and 200 million light-years wide.

Levy 332P   UGC371
Galaxy near Alpheratz
First seen: January 17, 2005
Position (2000.0): α 00 37.4   δ +29 09
Magnitude: 15.1
Distance: 236 million light-years
Best seen: in fall; needs dark sky
Classification: Scd

THE GREAT STRATUM

In 1936 Clyde Tombaugh detected one of the first known superclusters. Stretching from Pisces through Andromeda to Perseus, this vast array has recently been found to be associated with a second supercluster in the constellations of Ursa Major and Lynx. It is possible that the two groups are really joined, connected gravitationally in some way. If that is true, this "megacluster" stretches from horizon to horizon—across half the sky.

UGC371 represents the west "lobe," as Clyde Tombaugh called it, of his "Great Stratum" of galaxies. It is a faint, nondescript edge-on spiral, about 15 magnitude.

Levy 333P   PGC2218765 and PGC2218178
Galaxies near Algol
First seen: January 17, 2005
Position (2000.0): α 03 14.4   δ +43 21; α 03 14.4   δ +43 20
Magnitudes: 15.5 and 15.0
Distances: more than 200 million light-years
Best seen: in fall; needs dark sky
Classifications: E; Sa

> The human mind is so limited that it cannot take in all parts
> of a subject; so that there may be objections raised against
> anything. There are objections against a plenum [a universe
> filled], and objections against a vacuum. Yet one of them
> must certainly be true.
> —James Boswell, *Boswell's London Journal*, 1762–1763

These galaxies represent the east or Perseus "lobe" of Clyde
Tombaugh's stratum. I photographed this field the night that
Comet Machholz happened to be passing by. They are also part
of a great story about how we came to understand the distribution of galaxies and clusters of galaxies in space.

With the opening of the Mount Wilson 100-inch telescope
in 1918, astronomers had the most fantastic tool in the history of astronomy, for that time, to study the distant expanses
of the Universe. Edwin Hubble took advantage of this telescope to survey random areas of the sky to see how distant
galaxies were distributed. Although his telescope was large—
its light gathering mirror more than eight feet wide—its field
of vision was narrow. Hubble's method was to sample small
areas around the sky; he concluded that the clusters of galaxies
were distributed evenly through space, except for, of course,
the "zone of avoidance" where the Milky Way blocks any view
of a galaxy located behind it.

Clyde Tombaugh's Great Stratum led him to another conclusion. Tombaugh's method was different; during his long search for distant planets, he took notes on the distribution of galaxies in each of his photographs. Instead of sampling sectors of the sky, Tombaugh simply photographed all of it, as seen from Flagstaff, Arizona. Galaxy clusters are not distributed evenly at all, he found, but are clumped together.

In the mid-1940s, Tombaugh and Hubble debated the issue in Hubble's office at the California Institute of Technology. Tombaugh explained how, on his photographic plates covering all of the sky visible from Flagstaff, he did not see an even distribution of galaxy clusters. He reported dense concentrations of galaxies on his films, especially the mighty "stratum." He also reported "voids," where he saw little but empty space where there should have been galaxies. Hubble didn't take the younger scientist seriously. He should have. A decade later, George Abell of Palomar Observatory, in a doctoral thesis using photographic plates from the new Schmidt camera at Palomar Observatory, confirmed Tombaugh's view of the Universe and went on to show that clusters of galaxies are clumped into vast superclusters.

In addition to these findings, astronomers have also confirmed Tombaugh's voids—where empty space stretches on virtually forever. Voids are not completely without matter, but contain much less matter than in normal space.

QUASARS

Levy 134   NGC 4319 and Markarian 205
Galaxy in Draco
First seen: October 14, 2002
Position (2000.0): α 12 21.7 δ +75 19
Magnitude: 12.0
Distances: NGC 4319 is 87 million light-years; Markarian 205 is 1 billion
    light-years

Best seen: in spring; requires dark sky
Classifications: NGC 4319 is SBb; Markarian 205 is QSO

Until the Hubble Space Telescope solved the mystery, NGC 4319 and Markarian 205 were an astronomical enigma. Originally catalogd by the Armenian astronomer Benik Markarian, Markarian 205 is a distant object, either a low-luminosity quasar or a high-luminosity Seyfert galaxy. Astronomer Halton Arp pointed out that a filament of light seems to join NGC 4319 and Markarian 205, giving the impression that the two objects are physically related. But the redshifts—the shifts of their spectra toward the red end of the spectrum, as in Hubble's way of determining distance—are greatly discordant. If the two objects are really related, then the redshift, one of the building blocks of modern cosmology, might not seem to work in this case. The Hubble images taken in 1993 have helped solve this mystery however: the two objects only happen to be on our line of sight; one is farther than the other.

OLBERS'S PARADOX AND THE FUTURE OF THE UNIVERSE

Markarian 205 is weak as a quasar, but it is still an incredibly energetic object. Quasars are active galaxies, or the active cores of distant galaxies. In a sense, they continue an old tale from two centuries ago, when physician Heinrich Olbers looked pensively at the night sky over Bremen, Germany, and asked himself why it was dark. There are a certain number of bright stars, spaced at random intervals against the dark background, and a much higher number of fainter stars. We all know this, and it is easy to imagine that with each increasing magnitude, there are far more stars. Olbers's question: Should not the stars continue multiplying the fainter they get? Then why is the entire sky not blindingly bright with stars? If space is infinite and filled with stars, he suggested, then the entire sky should be as bright as the surface of the Sun. Obviously there are only a finite number of stars in our galaxy, but in the current age of our understanding of the Universe, the superclusters of galaxies should be filling the night sky with blinding radiation.

The answer to this old riddle, it seems, lies in the age of the Universe. If the Universe were several orders of magnitude older than it is—hundreds of billions of years old—and if we could see all the energy that ever existed within it—radiation from supercluster upon supercluster—it would indeed blind us. But the Universe, at an age of 13.7 billion years, is not old enough for light from its most distant objects—the galaxies and quasars—to have reached us yet. Another factor lies in the expansion of the Universe. Because it expands, radiation from distant objects is redshifted and weakened as every photon suffers a loss of energy during its shift to the red.

As each day ends, and we watch the Sun set, the world is left to darkness. At such a time we can ponder how incredible it is that this darkness encompassed the Universe as it was about to be born and still encompasses much of it as it continues to expand.

But expand to what? Will it expand forever?

If the Universe is "open," then it does not have enough gravity to stop its expansion, and it will eventually end with dead stars in ghost galaxies. Eventually, stars in galaxies will plunge into central black holes, and atomic particles will decay. After more billions of years than we could count (one estimate is ten to the one-hundreth power!), even quarks will vanish, leaving a little light, some leptons, and some slowly evaporating black holes.

If there is sufficient mass, and therefore sufficient gravity, to slow the expansion to a full stop, then a "closed" Universe will hold still for less than a second and start a slow process of collapse. Redshifts will become blueshifts, and over billions of years, the superclusters of galaxies will close in on each other. As they get very close to each other, temperature and pressure will rise until all atoms, of all stars, of all galaxies will dissolve into their nuclei. As the squeezing strengthens, nuclei will dissolve into protons and neutrons. Eventually these nucleic particles will dissolve into a soup of quarks and leptons. After uncounted billions of years, Olbers's paradox will be reconciled in a direct way: every "square inch" of sky will be as bright as the Sun.

Levy 230   3C-273 Virginis
Quasar in Virgo
First seen: April 1976
Position (2000.0): α 12 29.1   δ +02 03
Magnitude: 12.0
Distance: 2.5 billion light-years
Best seen: in spring; requires dark sky
Classification: QSO

> Every phenomenon, and every person, is a microcosm of the whole pattern of the universe, according to [Jung's] idea.
> —Tom Wolfe, *The Electric Kool-Aid Acid Test*, 1968

On the night of March 17, 1977, I was skywatching from a campground at the beautiful Organ Pipe Cactus National Monument on the border of Mexico and southwestern Arizona. That was the night I used a star chart supplied by the American Association of Variable Star Observers to get my first view of a quasar, 3C 273 Virginis—a very faint star through a 6-inch telescope, but one of the most fascinating things I have ever seen. Looking 2.5 billion light-years into space that night, I saw light from one of the most monstrous objects ever detected.

Had astronomers not had access to two different types of telescopes, optical and radio, the true nature of quasars might still elude us. In 1960 radio telescopes had revealed a strange source called 3C 48 that seemed to coincide with a faint blue star. It was the first star (other than the Sun) that seemed to emit radio waves. Some time later, another bright radio source, 3C 273, was blocked or occulted by the Moon three times within a period of several months. As the Moon passed in front of the radio source, the far-off object's "noise" was cut off abruptly. Because we know the position of the Moon precisely at any given time, these events allowed astronomers to determine the exact position of the radio source. In 1963 Maarten Schmidt turned the 200-inch on Palomar Mountain to the position that had been derived by the lunar occultations. He found a fairly bright star that was accompanied by a fainter jet of light. Not quite stars, these objects were given the name quasi-stellar radio sources. (This name was later shortened to quasars, and in our age of acronyms, quasi-stellar objects or QSOs.) Schmidt obtained a series of spectra of 3C 273 and realized that, if its redshift were interpreted correctly, that 13 magnitude star with its jet was billions of light-years away.

Despite its distance, 3C 273 is so bright that it can be spotted through a small telescope—if you know just where to look, and, with the aid of an atlas or computer program, you can distinguish it from the surrounding stars. Large telescopes reveal much more detail, including the existence of the aforementioned jet of material racing out of the quasar's center. A quasar is an energy machine so vast that it can create black holes simply from the collapse of clouds. A quasar's heart is not very much larger than our solar system. The more matter it contains, the more energy it releases. It is a powerful energy source like a supermassive black hole, which swiftly pushes electrons out to high speeds as they spiral outward. The electrons emit what we call synchrotron radiation (which we discussed in chapter 7 with regard to neutron stars). Surrounding it are rich clouds of gas that have erupted out of the center in times past.

Since 1979 astronomers have detected objects ever farther away in space, ever farther backward in time. An arc or primodial light that forms almost a complete circle was reported in 2005 by Rémi Cabanac and his team using Chile's Very Large Telescope (VLT). The lensing galaxy is 8 billion light years from Earth, but the remote object is 12 billion light years away, the farthest located so far. We are seeing this distant object as it appeared when the Universe was a little more than a tenth of its present age.[2]

Levy 337   Double Quasar Q0957+56A/B
Quasar in Ursa Major
First photographed: February 8, 2005
First seen visually: Valentine's Day morning 2005
Position (2000.0): α 10 01 21.05   δ +55 53 56.5
Magnitude: 15.5
Distance: 8.5 billion light-years
Classification: QSO

Tyger! Tyger! burning bright
In the forests of the night,
What immortal hand or eye
Could frame thy fearful symmetry?

In what distant deeps or skies
Burnt the fire of thine eyes?
On what wings dare he aspire?
What the hand dare seize the fire?

And what shoulder, and what art,
Could twist the sinews of thy heart?
And when thy heart began to beat,
What dread hand? and what dread feet?

What the hammer? what the chain?
In what furnace was thy brain?
What the anvil? what dread grasp
Dare its deadly terrors clasp?

When the stars threw down their spears,
And watered heaven with their tears,
Did he smile his work to see?
Did he who made the Lamb make thee?

Tyger! Tyger! burning bright
In the forests of the night,
What immortal hand or eye
Dare frame thy fearful symmetry?
                              —William Blake, "The Tyger," 1794

Levy 337,
Double Quasar
in Ursa Major.
Tom Glinos
Photograph.

Only a few days after my long journey from Canada to Arizona—a journey intended to find a clear sky for comet searching—I joined Dan Brocious, public affairs officer for the Smithsonian Institution's observatory. We drove up a long, single-lane dirt road to the mountaintop, where we saw a collection of telescopes, topped by the brand-new six-mirror Multiple Mirror Telescope. The warm September afternoon was memorable, but what really inspired me was the look at the new technology that Dan showed us, as well as a major discovery that was made with it.

The technology was a small computer chip smaller than a postage stamp. "Ordinary film captures less than half of the photons that strike it," he explained. "This 'CCD' chip is about 98 percent efficient," meaning that virtually all the light that strikes it gets recorded. He then told of one of the first discoveries using this new technology: a double quasar and an intervening galaxy that acts as a gravitational lens. The quasars were originally identified on a Palomar Sky Survey photograph in the late 1970s. Then, with the power of the then-new Multiple Mirror Telescope south of Tucson, Dennis Walsh, Bob Carswell, and Ray Weymann found that the quasars' redshifts were virtually identical. Could both images be of the same object? With the power of the CCD, they found the faint galaxy between the two images. The galaxy and the quasar were playing out an important part of Einstein's general theory of relativity—that gravitational sources could bend light.[3] The gravity of the galaxy was acting as a lens, bending the light of a more distant quasar so that it appeared double. This is the first discovery of the bending of light by gravitation since Arthur Stanley Eddington measured the deflection of the light of stars in the Hyades star cluster during the 1919 total solar eclipse.

Since 1979 astronomers have identified even more exotic examples of gravitational lenses. One galaxy is so elegantly positioned that the quasar far behind it is broken up into four images we now call Einstein's Cross. Abell 2228 is another example. One of the brightest galaxies in this cluster is lensing the light of far more distant galaxies behind it, giving them the appearance of long arcs of light. The distant galaxy, so dim that only the light of a world-class telescope can reveal it, was used as a telescope to allow us to see a quasar even more distant! What I didn't realize that day was that twenty-five years later I'd take a picture of that very quasar with my own backyard telescope.

Nor did I think that anyone would ever see that quasar visually. On the night of June 6, 1988, Bob Bunge and Brent Archinal made what is probably the first visual observation of a distant object—with the help of a telescope *and* a gravitational lens. One of the best visual telescopes in the world, the Richland Astronomical Society's 31-inch f/7 reflector, is certainly capable of revealing such a sight—I have looked through that telescope many times. They found the quasar, but because it was low in the sky they could not resolve it into its two images. On a later night, Bob used a 12 mm eyepiece at a power of 460 to split the quasar. Steve O'Meara also reported seeing it from the Texas Star Party and showed it to Pluto discoverer Clyde Tombaugh. O'Meara also saw it, but didn't split it, using a 7-inch refractor.[4]

A VALENTINE'S NIGHT TO REMEMBER

With all these experienced skywatchers reporting success trying to observe the Ursa Major quasar visually, I thought that I should try it myself. On Valentine's Day morning 2005, Rolf and Linda Meier and Wendee and I began an attempt to detect the quasar visually. Linda is an experienced variable star observer, Rolf a veteran of four comet discoveries. We had to wait for clouds to pass by that evening. Because the galaxy NGC 3079 almost points to the quasar, we all had an easy time at least seeing the quasar's field. Seeing the QSO itself was a challenge in Miranda (my 16-inch f/5 reflector) at 156 power, but once we got the pattern of stars that surrounded it right, the quasar appeared, looking like a very faint star.

We then went to Tom Glinos's RC Optical Systems 25-inch f/8 Ritchey-Chretien telescope. We inserted a 7 mm eyepiece in an attempt to see the effect of the gravitational lens. This combination yielded a comfortably high magnification of 726 power to try to split the lensed object, which has a separation of 6.3 arcseconds. When an object is near the limit of visibility, it doesn't appear constantly but tends to appear and disappear in brief glimpses, depending on the steadiness of the atmosphere above us and on the human eye-brain combination, trying to make sense of the faint detail. The quasar's double image appeared and disappeared many times, whenever the seeing permitted it. For me, whenever the quasar did appear, it always gave the appearance of two closely spaced but clearly separate objects—two lensed images of the same distant quasistellar object; two high-powered beams of light from the edge of the known universe.

It took about 8.5 billion years for the light of this QSO to travel through space to human eyes. Without the help of the intervening galaxy, there would have been no chance of seeing it. The galaxy's gravity split the optical path into two, allowing a view of two images of the quasar, each one magnified intensely thanks to the gravitational lens. The celestial couple we saw on Valentine's morning was in fact two identical images of William Blake's celestial tiger burning bright, its two eyes gazing at us in fearful symmetry.

It was a night we will not soon forget; only it was superceded just three weeks later! On March 9, 2005, I saw the quasar as two distinct images through Miranda, my own 16-inch reflector, using a magnification of 581 under a very steady sky. On May 8 Wendee and I saw the QSO again through the 25-inch; it is a difficult object.

A FINAL PUZZLE

There is one more intriguing possibility that connects further this last object of our survey to the stars and worlds at its start. The intervening galaxy—the lens—might have used the distant quasar to reveal a distant planet only three times larger than Earth!

When we haven't seen planets this small even around stars in our own galaxy, this sounds somewhat exotic. However it is consistent with the following observation. In 1996 Rudolph E. Schild and his team were observing the double quasar as revealed through the eye of the lensing galaxy. The way the galaxy works as a lens is that when both lobes, or lensed images, of the quasar that we see fluctuate, then the cause of the variation lies in the quasar. However, Schild's team found a variation of light, lasting a brief period of days—in just the A lobe and not in the B lobe. If a planet about three times the mass of Earth happened to pass directly in front of the quasar's distant point of light, it would produce exactly the effect that the team observed.

A FINAL THOUGHT

From grains of dust to a gravitational lens: We began our list with the Gegenschein, a collection of dust that barely qualifies as something that exists, and yet it lights up a large area of sky. We end it with light from one of the most intriguing sights in the Universe—a quasar made visible by a galaxy acting as a gravitational lens—yet it lights up so little sky it is barely visible to our telescopes. Both the Gegenschein and the quasar, as well as a host of objects in between, have called to me in the course of my long search of the night sky.

We can certainly imagine the millions of miles of distance within our solar system, and maybe even the light-years of distance to the closer stars. But it is hopeless to comprehend the vastness of space represented by the objects in this chapter. The quasar in Ursa Major is eight billion light-years away. How can we dream of such a distance? The numbers are so vast we cannot fathom them.

We can wonder, however. In the same field of the Ursa Major quasar is a galaxy, NGC 3079. It is 66 million light-years away, and it is a spiral not unlike our own Milky Way, except for some off activity in its northern arm. As we look at NGC 3079 and the nearby quasar, we have the intelligence to pose questions about the nature of life elsewhere in the Universe. Is it possible that just one of NGC 3079's stars has a planet with life? We do not have the intelligence to visualize a quasar 8 billion light-years away, but we do have the wherewithal to picture it, to ask questions about it, and to try to understand its nature. In fact, as large as the Universe is and of all the things it contains, the most complex item we have so far encountered lies not in that quasar but in the human brain. Far more complex than the gravity that forms a galaxy, a star, and a world, the human brain allows us to lie awake at night, dream about things that are so far away, and then go outdoors and see them through a telescope of our making.

Levy 342P, Arp 65, an absolutely magnificent cluster of galaxies. NGC 91, the spiral with the arms stretching on and on like the Energizer bunny, is 258 million light-years away.

PART 3

THE FULL LEVY LIST

... like starry light
Which sparckling on the silent waves, does
seem more bright.
  —Edmund Spenser, *Faerie Queene* 2:78, 1590

Up to now, this book has been but an introduction to the best of the objects I have seen in the night sky. Now, I offer two versions of the full catalog of the deep sky objects that have brightened my nights over forty years of chasing after comets. The first list is in order of when each object was added. Note that for objects already listed in earlier chapters, the date is different. In those chapters, I listed the date I first saw the object, often, for the brighter Messiers, well

before encountering them while comet hunting. In the list that follows, that date refers to my first sighting *while comet hunting*.

This list is kept updated at the following Web site: http://www.jarnac.org. Click on "Levy List of Objects."

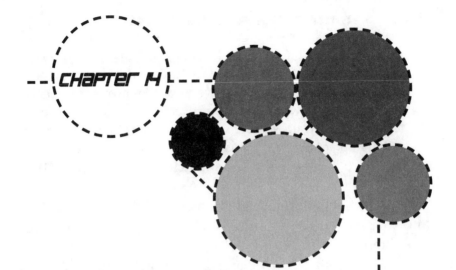

CHAPTER 14

DAVID LEVY'S CATALOG OF COMET MASQUERADERS AND OTHER OBJECTS

Of the president of a certain agricultural college on his first trip to Europe, this story is told. He was leaning pensively over the ship's rail when he was offered a penny for his thoughts. Waving his arm over the vasty deep which has touched the imagination more often and deeply than anything else in nature except, perhaps, the starry heavens, he replied: "You know, I was just thinking what a pity it is that all this can't be put down in alfalfa." There are, no doubt, others

267

who never see a bird without thinking of the pot, or a wood without leaping happily forward to the day when it will make newsprint, or who never observe any stretch of wild open country without wishing it were covered with skyscrapers or factories or bungaloes. More souls for heaven? They will be, I think, ill prepared for it.

—Joseph Wood Krutch, *The Desert Year*, 1951

This is a catalog of objects that have caused me to stop, look, and wonder during my thirty-seven years of comet hunting. The asterisked objects, like 2, 14, 28, 36, 43, 45, and 58, and especially 37, 43, 78, 161, and 211, clearly "masquerade" as comets, especially when they are low in the sky, which is where I usually spot them. The rest of the list contains different types of interesting objects—like no. 61, the Cetus Ring—that have stopped my search over some 2,758 hours with my eye at the eyepiece (as of July 2002).

Although the list is generally in chronological order of when I first spotted an object, several objects near the end were added later. Names in quotations are names that Wendee or I applied to specific objects.

In the following two lists, note that "L1" is short for "Levy 1" and so on.

No.	Other Design.	Spotted on	α (2000.0)	δ	Comment
L1	**NGC 1931**	1/1/1966	05 31.4	+34 15	Nebula
*L2	**NGC 5457 M101**	7/4/1966	14 03.2	+54 21	Galaxy
L3	**NGC 6341 M92**	7/5/1966	17 17.1	+43 08	Globular cluster
L4	**NGC 6254 M10**	7/13/1966	16 57.1	−04 06	Globular cluster
L5	**NGC 5676**	7/13/1966	14 32.8	+49 28	Galaxy
L6	**NGC 6229**	7/14/1966	16 47.0	+47 32	Globular cluster
L7	**NGC 5055 M63**	7/14/1966	13 15.8	+42 02	Sunflower Galaxy
L8	**NGC 1624**	7/16/1966	04 40.4	+50 27	Cluster with nebulosity; rich background
L9	**NGC 2403**	7/16/1966	07 36.9	+65 36	Galaxy
L10	**NGC 2655**	7/20/1966	08 55.6	+78 13	Galaxy
L11	**NGC 4605**	8/5/1966	12 40.0	+61 37	Galaxy

L12	NGC 7078 M15	8/23/1966	21 30.0	+12 10	Globular cluster
L13	NGC 6720 M57	9/7/1966	18 53.6	+33 02	Ring Nebula
*L14	NGC 2068 M78	9/10/1966	05 46.7	+00 03	Comet-like nebula; on 8/16/2002, I observed that the complex includes NGC 2071 around the star to the north
L15	NGC 4826 M64	9/19/1966	12 56.7	+21 41	Black Eye Galaxy
L16	NGC 3627 M66	3/3/1967	11 20.2	+12 59	Galaxy
L17	NGC 6853 M27	3/3/1967	19 59.6	+22 43	Dumbell Planetary
L18	NGC 2392	5/5/1967	07 29.2	+20 55	Clown face
L19	NGC 6207	5/6/1967	16 43.1	+36 50	Galaxy
L20	NGC 6402 M14	5/6/1967	17 37.6	−03 15	Globular cluster
L21	NGC 6838 M71	1967	19 53.8	+18 47	Globular cluster
L22	NGC 3034 M82	12/24/1967	09 55.8	+69 41	Galaxy
L23	NGC 5024 M53	5/5/1968	13 12.9	+18 10	Globular cluster
L24	NGC 3031 M81	11/27/1968	09 55.6	+69 04	Galaxy
L25	NGC 1904 M79	1/22/1969	05 24.5	−24 33	Globular cluster
L26	NGC 4258 M106	6/25/1970	12 19.0	+47 18	Pear-shaped galaxy
L27	NGC 5377	8/6/1970	13 56.3	+47 14	Galaxy
*L28	NGC 5473	8/6/1970	14 04.7	+54 54	Galaxy
L29	NGC 5474	8/6/1970	14 05.0	+53 40	Galaxy
L30	NGC 5866 M102?	8/8/1970	15 06.5	+55 46	Galaxy
L31	NGC 514	8/29/1970	01 24.1	+12 55	Galaxy
L32	NGC 7331	9/3/1970	22 37.1	+34 25	Galaxy
L33	NGC 1952 M1	8/29/1970	05 34.5	+22 01	Supernova remnant
L34	NGC 488	9/12/1970	01 21.8	+05 15	Elongated galaxy
L35	NGC 2420	9/12/1970	07 38.5	+21 34	Cluster
*L36	NGC 6364	1/11/1983	17 24.5	+29 24	Comet-like galaxy
*L37	NGC 3055	1981	09 55.3	+04 16	Wendee's galaxy
L38	NGC 7753	4/28/1985	23 47.1	+29 20	Galaxy; broad toward center
L39	NGC 6791		19 20.7	+37 51	Faint open cluster; dense
*L40	NGC 5466		14 05.5	+28 32	Globular cluster in Boötes; comet-like but mottled (M=9.1; D=11')
L41	NGC 3810		11 41.0	+11 28	Round galaxy; bright core; diffuse
L42	NGC 4340		12 23.6	+16 43	Galaxy in Coma Berenices
*L43	NGC 7793	1/23/1989			Like a diffuse comet
L44	NGC 1333	7/2/2000	03 29.2	+31 25	Weird reflection nebula
*L45	NGC 2188	12/17/2001	06 10.0	−34 06	Comet-tail galaxy
L46	NGC 1851	12/17/2001	05 14.0	−40 03	Oval globular cluster
L47	NGC 1679	12/17/2001	04 50.0	−31 59	Round galaxy
L48	NGC 7582	12/17/2001	23 18.4	−42 22	Galaxy
L49	NGC 1941	07/22/2001	04 03.4	+51 19	Reflection nebula; near a star
L50	NGC 1579	07/22/2001	04 30.2	+35 16	Reflection nebula

The man of science like the man on the street has to face hardheaded facts that cannot be blinked and explain them as best he can. There may be, it is true, some questions which science cannot answer—at present . . .

—James Joyce, *Ulysses*, 1922

L51	**Arp 321**	1985, 2001	09 38.9	−04 52	Galaxies in Hydra
L52	**NGC 3115**		10 05.2	−07 43	Spindle Galaxy
L53	**NGC 752**	07/04/2002	01 57.8	+37 41	Open cluster
L54	**NGC 5128**	12/19/2001	13 25.5	−43 01	Galaxy
L55	**NGC 7664**	06/09/2002	23 26.6	+25 04	Galaxy
L56	**NGC 270**	06/09/2002	00 50.6	−08 39	Galaxy
L57	**NGC 7723**	06/09/2002	23 38.8	−12 58	Galaxy in Aquarius
*L58	**NGC 772**	06/09/2002	01 59.3	+19 01	Round, comet-like galaxy with bright core
L59	**NGC 524**	06/09/2002	01 24.8	+09 32	Round galaxy
L60	**NGC 628 M74**	06/09/2002	01 36.6	+15 47	Diffuse galaxy
L61	**NGC 246**	06/18/2002	00 47.0	−11 53	Planetary nebula
L62	**NGC 898**	06/19/2002	02 23.3	+41 57	Very elongated and faint galaxy in Andromeda
L63P	**NGC 4656**	07/06/2002	12 44.0	+32 10	"Hummingbird" Galaxy
L64	**NGC 404**	07/04/2002	01 09.4	+35 43	Round galaxy near Beta Andromedae; bright core
L65	**NGC 4374 M84**	07/03/2002	12 25.1	+12 53	Galaxy
L66	**NGC 4406 M86**	07/03/2002	12 26.2	+12 57	Galaxy
L67	**NGC 4594 M104**	07/03/2002	12 40.0	−11 37	Sombrero Galaxy
L68	**NGC 1023**	07/04/2002	02 40.4	+39 04	Very elongated galaxy in Perseus
L69P	**LW J2204.3+4508**	01/02/2000	22 04.3	+45 08	"Wendee's Ring"
L70P	**LW J2108.8+0620**	12/25/2000	21 08.8	+06 20	"Equuleus S"
L71	**LW J2340.6+5618**	05/03/2001	23 40.6	+56 18	"Nanette's River"
L72	**V Hydrae**	11/22/1984	10 51.6	−21 15	Red star
L73	**TV Corvi**	03/23/1990	12 20.4	−18 27	Tombaugh's star
L74	**NGC 3623 M65**	3/3/1967	11 18.9	+13 05	Glaxy
L75	**NGC 3372**	2/9/2000	10 45.1	−59 41	Eta Carinae Nebula
L76	**NGC 6618 M17**		18 20.7	−16 10	Nebula
L77	**NGC 2237**	09/04/2002	06 32.3	+05 03	The Rosette complex: cluster 2244; nebulae 2237, 2238, 2239, and 2246
*L78	**NGC 2261**		06 39.2	+08 44	Hubble's Nebula
L79	**NGC 3628**		11 20.3	+13 36	Edge-on galaxy
*L80	**NGC 185**		00 39.0	+48 20	Galaxy
*L81	**NGC 147**		00 33.2	+48 30	Galaxy
L82	**NGC 5634**		14 29.6	−05 59	Globular cluster; stars on either side

*L83	**NGC 5638**		14 29.7	+03 14	Comet-like galaxy
L84	**NGC 6712**		18 53.1	–08 42	Faint globular
L85	**IC 1396**	1966	21 39.1	+57 30	Open cluster
L86	**NGC 224 M31**	07/04/1966	00 42.7	+41 16	Andromeda Galaxy
L87	**NGC 5194 M51**	07/13/1966	13 29.9	+47 12	Whirlpool Galaxy
L88	**NGC 3587M97**	07/13/1966	11 14.8	+55 01	Owl Nebula
L89	**NGC 3556 M108**		11 11.5	+55 40	Galaxy
L90	**NGC 3992 M109**		11 57.6	+53 23	Galaxy
L91	**NGC 598 M33**	08/14/1966	01 33.9	+30 39	Low surface brightness galaxy
L92	**NGC 253**	10/1979	00 47.6	–25 17	Galaxy
L93	**NGC 2683**	10/26/1979	08 52.7	+33 25	Comet-like galaxy; very elongated, dusty, bright core
L94	**NGC 6723**	11/05/1979	18 59.6	–36 38	Globular cluster; highly resolved; near L295
L95	**NGC 1999**	11/15/1979, 04/09/2000	05 36.5	–06 42	Diffuse nebula
L96	**NGC 2681**	11/15/1979	08 53.5	+51 19	Galaxy
L97	**NGC 5139**	05/09/1980	13 26.8	–47 29	Omega Centauri
L98	**NGC 6709**	11/13/1984	18 51.5	+10 21	Open cluster
*L99	**NGC 949**	06/15/1997	02 30.8	+37 08	Very cometary galaxy in Triangulum
L100	**â Persei, Algol**	03/13/1997	03 08.1	+40 58	Variable star

Meditations of evolution increasingly vaster: . . . of Sirius (Alpha in Canis Major) 10 lightyears (57,000,000,000,000 miles) distant . . . of moribund and nascent new stars such as Nova in 1901: of our solar system plunging towards the constellation of Hercules: of the parallax or parallactic drift of socalled fixed stars, in reality evermoving from immeasurably remote eons to infinitely remote futures in comparison with which the years, threescore and ten, of allotted life formed a parenthesis of infinitesimal brevity.

—Joyce, *Ulysses*, 1922

L101	**NGC 3766**	02/08/2000	11 36.1	–61 37	Open cluster
L102	**IC2602**	02/08/2000	10 43.2	–64 24	Big nearby open cluster in Carina
L103	**NGC 3621**	05/06/2000	11 18.3	–32 49	"Frame Galaxy"
L104	**NGC 104 and 121**	06/18/2001	00 24.1	–72 05	Very bright globular
			00 26.7	–71 32	Faint globular in SMC
L105	**NGC 362**	06/18/2001	01 03.2	–70 51	Globular cluster in Tucana; brilliant

L106	**NGC 4038/4039**	2001	12 01.9	−19 52	The Antennae or Ring Tail; colliding galaxies in Corvus
L107	**NGC 4361**	2001	12 24.5	−18 48	Planetary nebula
L108	**Struve 1604 Corvi**	06/30/2002	12 09.5	−11 52	Triple star
L109	**NGC 936**	07/09/2002	02 27.6	−01 09	Galaxy; bright core
L110	**NGC 6760**	07/18/2002	19 11.2	+01 02	Aquila globular cluster
L111	**NGC 1068 M77**	07/18/2002	02 42.7	−00 01	Cetus galaxy; very bright core
L112	**NGC 7023**	08/13/2002	21 00.5	+68 10	Most unusual-appearing nebula with dust; star at edge
L113	**TU Geminorum**	08/15/2002	06 10.9	+26 01	Bright semiregular red variable star; range 6.2–8.6
*L114	**NGC 1637**	08/16/2002	04 41.5	−02 51	Comet-like round galaxy (M=10.9; D=3.9')
*L115	**NGC 1788**	08/16/2002	05 06.9	−03 21	Reflection nebula with "tail"; 8 arcmin diameter
L116	**NGC 2158**	08/17/2002	06 07.5	+24 06	Compact open cluster near M35; looks nebulous at low power
L117	**NGC 6093 M80**	08/25/2002	16 17.0	−22 59	Compact globular cluster; looked fuzzy when first sighted in 1980s
L118	**Sagittarius Star Cloud**		18 03.4	−27 54	Stars and dust lanes
L119	**NGC 6603 M24**	08/26/2002	18 16.9	−18 29	Small Sgr star cloud
L120	**NGC 6451**	08/26/2002	17 50.7	−30 13	Small Scorpius OC; Galactic Ctr:
			17 45.6	−28 56	~2° SE from galactic center
L121	**NGC 5846**	08/26/2002	15 06.4	+01 36	Virgo; roundish galaxy
L122	**NGC 6826**	08/29/2002	19 44.8	+50 31	Blinking planetary nebula
L123	**NGC 1600**	09/04/2002	04 31.7	−05 05	Eridanus round galaxy; diffuse
L124	**NGC 2174**	09/04/2002	06 09.7	+20 30	Large field of dust; 38 arcmin; barely noticeable but large area of brightening
L125	**NGC 2023**	09/04/2002	05 41.6	−02 14	Complex includes IC434 Barnard 33, the Horsehead
L126	**NGC 2359**	10/01/2002	07 18.6	−13 12	Thor's Helmet
L127	**16/17 Draconis**	10/05/2002	16 36.2	+52 55	Thinks it's a copy of Epsilon Lyrae, but only one star is binary
L128P	**IC5020 PGC64845**	10/07/2002	20 30.6	−33 29	Photographic addition; galaxy with a line of foreground stars; part looks like question mark
L129	**UGC5373**	10/08/2002	10 00.0	+05 20	Sextans B; local group member; extremely wide, about 1/4°

L130	**NGC 3198**	10/08/2002	10 19.9	+45 33	M1= 10.3, 8' diameter; very elongated galaxy
*L131	**NGC 2964-2968**	10/09/2002	09 42.9	+31 51	NGC 2964 brighter than 2968
L132	**NGC 3432**	10/09/2002	10 52.5	+36 37	Lmi galaxy; (M1= 11.3; D= 6.6) elongated with star at S side and star at W end.
*L133	**NGC 3070**	10/11/2002	09 58.0	+10 22	Leo round galaxy; (M1= 12.3; D=1.3) diffuse; faint companion 3069 not seen
L134	**NGC 4319**	10/14/2002	12 21.7	+75 19	Galaxy
L135	**NGC 4256**	10/14/2002	12 18.7	+65 54	Draco edge-on galaxy with bright core (M=11.9; D= 4.2)
*L136	**NGC 3738**	10/14/2002	11 35.8	+54 31	Galaxy near M97
*L137	**NGC 3718**	10/14/2002	11 32.6	+53 04	U Ma elongated galaxy
*L138	**NGC 3953**	10/14/2002	11 53.8	+52 20	See L90 M109, similar galaxy; very elongated with bright core
L139	**NGC 4449**	10/18/2002	12 28.2	+44 06	Galaxy, bright core
L140	**NGC 4485**	10/18/2002	12 30.5	+41 42	Elongated galaxy
L141	**NGC 4565**	10/18/2002	12 36.3	+25 59	Beautiful edge-on galaxy
*L142	**NGC 4274**	10/18/1999	12 19.8	+29 37	Elongated galaxy; bright core, 4278 and 4314 nearby, but neither are as cometary
*L143	**NGC 4559**	10/18/2002	12 36.0	+27 58	Elongated galaxy
*L144	**NGC 4501 M88**	10/18/2002	12 32.0	+14 25	Very elongated galaxy; bright core
*L145	**NGC 4473**	10/18/2002	12 29.8	+13 26	Very elongated galaxy with bright core
*L146	**NGC 4472 M49**	10/18/2002	12 29.8	+08 00	Virgo round galaxy; bright core
L147	**NGC 6638**	10/24/2002	18 30.9	−25 30	Globular cluster, added by Dean Koenig
L148	**R Leporis**	10/31/2002	04 59.6	−14 48	Red star
*L149	**NGC 6553**	11/04/2002	18 09.3	−25 54	Globular, faint and large
L150	**NGC 6523 M8**	11/04/2002	18 03.8	−24 23	Nebula

> O God, I could be bounded in a nutshell and count myself a king of infinite space, were it not that I have bad dreams.
> —William Shakespeare, *Hamlet* 2.2.253–55, circa 1600

L151	**NGC 6611 M16**	11/04/2002	18 18.8	−13 47	Eagle Nebula with cluster
L152	**NGC 6514 M20**	11/04/2002	18 02.3	−23 02	Trifid Nebula
L153	**NGC 7293**	9/11/1982	22 29.6	−20 48	Helix Nebula; looks ghostlike

L154	**NGC 7009**	11/04/2002	21 04.2	−11 22	Saturn Nebula; blue!
L155	**NGC 6981 M72**	11/04/2002	20 53.5	−12 32	Globular
L156	**NGC 6934**	11/04/2002	20 34.2	+07 24	Globular cluster
L157	**LW J1948.2+3743**	11/04/2002	19 48.2	+37 43	The Cane
L158	**"Castor Cluster"**	11/04/2002			
	IC2196		07 34.1	+31 24	Galaxy
	IC2197		07 34.3	+31 24	Galaxy
	IC2194		07 33.7	+31 19	Galaxy
L159	**NGC 2264**	11/04/2002	06 41.1	+09 53	Cluster with nebulosity
L160	**NGC 2254**	11/04/2002	06 36.0	+07 40	Open cluster and star chain
*L161	**NGC 2245**	11/04/2002	06 32.7	+10 10	Comet-like bright nebula, looks like Hubble's Nebula
L162	**NGC 2252**	11/04/2002	06 35.0	+05 23	Open cluster, looks like a rope of stars
*L163	**NGC 2775**	10/20/1979	09 10.3	+07 02	Round galaxy
*L164	**NGC 3486**	11/04/2002	11 00.4	+28 58	Round galaxy
*L165	**NGC 3245**	11/04/2002	10 27.3	+28 30	Elongated galaxy
*L166	**NGC 3344**	11/04/2002	10 43.5	+24 55	Round galaxy, bright core, star nearby (M=9.9; D=6.9); Don Machholz quotes Peltier as saying this looks like a comet; George Alcock echoes—its cometary appearance is enhanced by the nearby star
L167	**NGC 3310**	11/04/2002	10 38.7	+53 30	Round galaxy
L168	**NGC 3242**	11/04/2002	10 24.8	−18 38	Ghost of Jupiter planetary nebula
*L169	**NGC 2986**	11/04/2002	09 44.3	−21 17	Round galaxy; bright core
L170	**NGC 4651**	11/04/2002	12 43.7	+16 24	Round galaxy; bright core
L171	**NGC 4450**	11/04/2002	12 28.5	+17 05	Elongated galaxy; bright core
*L172	**NGC 4689**	11/04/2002	12 47.8	+13 46	Elongated galaxy
*L173	**NGC 4548 M91**	11/04/2002	12 35.4	+14 30	Elongated galaxy; bright core
*L174	**NGC 4649 M60**	11/04/2002	12 43.7	+11 33	Virgo galaxy; fainter one nearby
L175	**NGC 4486 M87**	11/04/2002	12 30.8	+12 24	Virgo elliptical galaxy
*L176	**NGC 4579 M58**	11/04/2002	12 37.7	+11 49	Virgo galaxy
L177	**NGC 4552 M89**	11/04/2002	12 35.7	+12 33	Virgo galaxy
*L178	**NGC 4596**	11/04/2002	12 39.9	+10 11	Elongated galaxy; bright core
*L179	**NGC 4535**	11/04/2002	12 34.3	+08 12	Round galaxy; bright core
*L180	**NGC 4303 M61**	11/04/2002	12 21.9	+04 28	Spiral galaxy; low surface brightness
*L181	**NGC 3887**	11/04/2002	11 47.1	−16 51	Round galaxy
L182	**NGC 4636**	11/10/2002	12 42.8	+02 41	Round galaxy; bright core
*L183	**NGC 4818**	11/10/2002	12 56.8	−08 31	Very elongated galaxy; barely visible in 16-inch

*L184	**NGC 5147**	11/10/2002	13 26.3	+02 06	Round galaxy; faint
*L185	**NGC 5248**	11/10/2002	13 37.5	+08 53	Round galaxy
*L186	**NGC 5371**	11/11/2002	13 55.7	+40 28	Round galaxy; beautiful, tightly wound spiral in photographs; stars in field
*L187	**NGC 5020**	11/11/2002	13 12.6	+12 36	Elongated galaxy; bright core
*L188	**NGC 4591**	11/11/2002	12 39.3	+06 01	Virgo elongated galaxy; faint through 16 mm eyepiece
L189	**NGC 5127**	11/12/2002	13 23.8	+31 34	Round galaxy; bright core; faint
L190	**NGC 4956**	11/12/2002	13 05.1	+35 11	Round galaxy
L191	**NGC 4772**	11/12/2002	12 53.5	+02 10	Galaxy
*L192	**NGC 4536**	11/12/2002	12 34.5	+02 11	Very elongated galaxy
*L193	**NGC 4129**	11/12/2002	12 08.9	−09 02	Elongated galaxy; 16 mm eyepiece
L194	**NGC 281**	11/13/2002	00 52.8	+56 37	Open cluster with well-shaped nebulosity; found photographically via Schmidt camera
L195	**NGC 2419**	11/13/2002	07 38.1	+38 53	Shapley's Intergalactic Wanderer globular cluster
L196	**NGC 5694**	circa 1990	14 39.6	−26 32	Tombaugh's globular cluster
L197	**NGC 5907**	10/24/1982	15 15.9	+56 19	Very long galaxy
L198	**NGC 2437 M46**	03/15/1983	07 41.8	−14 49	Open cluster
	NGC 2438		07 41.8	−14 44	Planetary nebula in foreground of M46
L199	**NGC 4567/4568**	03/15/1983	12 36.5	+11 15	Siamese twins galaxies
L200	**Fornax cluster**	1983			Cluster of several galaxies in same field
	NGC 1380		03 36.5	−34 59	
	NGC 1399		03 38.5	−35 27	
	NGC 1404		03 38.9	−35 35	

Is it so much, and yet the morn not up?
See yonder where the 'shame-faced maiden comes
Into our sight, how gently doth shee slide,
Hiding her chaste cheeks like a modest Bride,
With a red vaile of blushes.

—Francis Beaumont and John Fletcher,
The Woman Hater 1.1.3–8, 1607, about the rising of Virgo
just before dawn on a December morning

*L201	**NGC 3865**	03/05/1989	11 44.9	−09 14	Galaxy; faint and diffuse
*L202	**NGC 5427**	03/05/1989	14 03.4	−06 02	Galaxy

*L203	NGC 5668	03/05/1989	14 33.4	+04 27	Galaxy
*L204	NGC 5850	03/05/1989	15 07.1	+01 33	Galaxy
*L205	NGC 6106	03/05/1989	16 18.8	+07 25	Galaxy
*L206	NGC 6118	03/05/1989	16 21.8	−02 17	Round galaxy; very large; low surface brightness
*L207	NGC 6384	03/05/1989	17 32.4	+07 04	Galaxy
*L208	NGC 6426	03/05/1989	1744.9	+03 00	Globular cluster
*L209	NGC 3049	03/06/1989	09 54.8	+09 16	Galaxy
*L210	NGC 4685	11/13/2002	12 47.1	+19 28	"Winking Galaxy"
*L211	NGC 4779	11/13/2002	12 53.8	+09 44	Round galaxy; 16 mm eyepiece.
*L212	NGC 4795	11/13/2002	12 55.0	+08 04	Round galaxy; 16 mm eyepiece
*L213	NGC 4623	11/13/2002	12 42.2	+07 41	Very elongated galaxy; 16 mm eyepiece.
*L214	NGC 4713	11/13/2002	12 50.0	+05 19	Elongated galaxy; 16 mm eyepiece.
*L215	NGC 4688	11/13/2002	12 47.8	+04 20	Very large, low surface brightness galaxy; looks like a diffuse comet; 16 mm eyepiece
L216	NGC 4590 M68	11/15/2002	12 39.5	−26 45	Globular cluster; diffuse at low altitude
L217	NGC 3923	11/15/2002	11 51.0	−28 48	Elongated galaxy; bright core
L218	NGC 3201	11/15/2002	10 17.6	−46 25	Globular cluster; large
L219	Hydra I cluster	11/17/2002			
	NGC 3309		10 36.6	−27 31	
	NGC 3311		10 36.7	−27 32	
	NGC 3312		10 37.0	−27 34	
	NGC 3314		10 37.4	−27 41	Actually two spirals, one directly in front of the other
	NGC 3316		10 37.6	−27 36	
L220	NGC 6910	12/17/2002	20 23.1	+40 47	Open cluster near Gamma Cygni; 12/25/2002—encountered visually
*L221	NGC 300	01/01/2003	00 54.9	−37 41	Very large, low surface brightness galaxy
*L222	NGC 289	01/01/2003	00 52.7	−31 12	Elongated galaxy
*L223	NGC 150	01/01/2003	00 34.3	−27 48	Elongated galaxy
L224	IC1830	01/01/2003	02 39.1	−27 27	Galaxy near a star
L225	NGC 1187	01/01/2003	03 02.6	−22 52	Round galaxy
L226	NGC 2613	01/01/2003	08 33.4	−22 58	Edge-on galaxy
L227	Haffner17	01/01/2003	07 51.6	−31 49	Open cluster; dim with faint stars
*L228	NGC 4490	01/01/2003	12 30.6	+41 38	Elongated galaxy
L229	SS Virginis	01/01/2003	12 25.2	+00 46	Red (Spectral class Ne or C5 3E) variable star

L230	3C273 Virginis	03/17/1977	12 29.1	+02 03	Quasar close to SS Virginis
*L231	NGC 5364	01/01/2003	13 56.2	+05 01	Elongated galaxy; 5363 nearby
*L232	NGC 5068	01/01/2003	13 18.9	–21 02	Very large galaxy; low surface brightness but bright core
L233	NGC 4414	10/18/1999	12 26.4	+31 13	Galaxy
L234	NGC 4214	10/18/1999	12 15.6	+36 20	Galaxy
L235	NGC 4244	10/18/1999	12 17.5	+37 49	Very elongated galaxy
L236	NGC 6181	01/12/2003	16 32.3	+19 50	Galaxy
L237	NGC 6960		20 45.7	+30 43	Veil Nebula, western segment; involved with star
	NGC 6992		20 56.4	+31 43	Veil Nebula, main eastern segment
	NGC 6995		20 57.1	+31 13	Veil Nebula, southeastern segment
*L238	NGC 578	01/21/2003	01 30.5	–22 40	Elongated galaxy
*L239	NGC 247	01/21/2003	00 47.1	–20 46	Very elongated galaxy; foreground star at south end
*L240	NGC 157	01/21/2003	00 34.8	–08 24	Amoeba Galaxy; Cetus elongated galaxy; strange shape; faint foreground star at north end; looks like an amoeba
L241	AA Ceti	01/21/2003	01 59.0	–22 55	Double star; separation 8.4 at 304 degrees; AA is eclipsing variable
*L242	NGC 1165	01/27/2003	02 58.7	–32 06	Very elongated galaxy; very low surface brightness visually through 16-inch; bright and obvious on image through 6-inch
L243	NGC 6440	02/08/2003	17 48.9	–20 22	Globular cluster; very close to NGC 6444 (L244 below)
L244	NGC 6445	02/08/2003	17 49.2	–20 01	Planetary nebula
L245	NGC 6781	03/05/2003	19 18.4	+06 33	Planetary nebula; diffuse and interesting
L246	V460 Cygni	04/27/2003	21 42.0	+35 31	Red variable star; irregular variation; range 5.4–7.4
L247	NGC 7217	04/27/2003	22 07.9	+31 22	Round galaxy
L248	δ Scorpii	05/2003	16 00.5	–22 38	Interesting variable star has been in a lengthy outburst
L249	NGC 7457	05/19/2003	23 01.0	+30 09	Elongated galaxy; bright core
*L250	NGC 6946	05/22/2003	20 34.8	+60 09	Galaxy

Not from the stars do I my judgment pluck,
And yet methinks I have astronomy;
But not to tell of good or evil luck
Of plagues, or dearths, or seasons' quality . . .
> —Shakespeare, "Sonnet 14," circa 1600,
> against judicial astrology

L251	NGC 7789	05/23/2003	23 57.0	+56 44	Open cluster; faint and dense
L252	65 Psc	05/23/2003	00 49.9	+27 43	Close double star; separation: 4.4"
L253	BC Andromedae	05/23/2003	23 01.0	+46 31	M7 III red star
L254	NGC 869/884	10/28/1962	02 19.0	+57 09	Open clusters
L255	NGC 5016	10/26/2003	13 12.1	+24 06	Round galaxy
L256	IC2367	10/26/2003	08 24.2	−18 46	Galaxy
*L257	NGC 4866	11/29/2003	12 59.5	+14 10	Elongated galaxy
L258	NGC 2477	11/29/2003	07 52.3	−38 33	Open cluster; dense, cometary appearance when near horizon
L259	NGC 4722	11/29/2003	12 51.5	−13 19	Round galaxy; double
L260	NGC 4519	11/29/2003?	12 33.5	+09 39	Galaxy
L261	NGC 5350	11/29/2003?	13 53.4	+40 22	Galaxy
L262	NGC 6814	12/17/2003?	19 42.7	−10 19	Round galaxy
L263	NGC 7006	12/17/2003	21 01.5	+16 11	Globular cluster
L264	NGC 5897	12/20/2003	15 17.4	−21 01	Globular cluster; loose and diffuse
L265	NGC 5746	12/24/2003	15 44.9	+01 57	Elongated galaxy; dusty; a good candidate for imaging
L266	NGC 5962	12/24/2003	15 36.5	+16 37	Round galaxy
*L267	NGC 5247	01/18/2004	13 38.1	−17 53	Spiral galaxy with bright core; very diffuse when low in sky
L268	NGC 5690	01/18/2004	14 37.7	+02 17	Virgo; long, very elongated galaxy; near a bright star
L269	NGC 5460	01/18/2004	14 07.6	−48 19	Centaurus open cluster with a beautiful curve of stars
L270	NGC 6535	01/25/2004	18 03.8	−00 18	Globular cluster; mottled
*L271	NGC 6287	01/25/2004	17 05.2	−22 42	Globular cluster; unresolved; comet-like when low in the sky
L272	NGC 262, Swift II, Markarian 348		00 48.8	+31 57	Very massive galaxy 1.3 million light-years in diameter
L273	NGC 1746	04/08/2004	05 03.6	23 49	Open cluster
L274	NGC 6541	05/15/2004	18 08.0	−43 42	Globular cluster; small but thick center that spreads out quickly

L275	TX Piscium	05/15/2004	23 46.5	+03 29	Red variable star; also called 19 Piscium
*L276	NGC 7814	06/12/2004	00 03.3	+16 09	Elongated galaxy; bright core
L277	NGC 7184	06/18/2004	22 02.7	−20 49	Very elongated galaxy; brightest in a group
L278	NGC 474	06/18/2004	01 20.3	+03 26	Elliptical galaxy; brightest in group; includes NGC 467 and NGC 470
L279	NGC 6522	05/27/2004	18 03.6	−30 02	Baade's Window; two faint
	NGC 6528		18 04.8	−30 03	globular clusters close to galactic center
*L280	NGC 7721	06/19/2004	23 38.8	−06 31	Elongated galaxy; soft, long, comet-like
L281	NGC 7727	06/19/2004	23 39.9	−12 18	Round galaxy with bright core; a blinking galaxy; Seyfert galaxy
L282	NGC 7314	06/19/2004	22 35.8	−26 03	Elongated galaxy with bright core; looked a bit like S-L 9 just after discovery
L283	U Camelopardalis	06/19/2004	03 41.7	+62 40	Red variable star with nearby blue star; period 419D; Range 7.6-8.8.
*L284	NGC 718	06/25/2004	01 53.2	+04 12	Round galaxy with bright core; faint when low in sky
L285	NGC 6101	07/14/2004	16 25.8	−72 12	Globular star cluster
L286	NGC 1360	07/14/2004	03 33.3	−25 51	Planetary nebula
L287	NGC 1365	07/14/2004	03 33.6	−36 08	Magnificent barred spiral galaxy
L288	NGC 1097	07/15/2004	02 46.3	−30 17	Spectacular barred spiral
L289	IC5148	07/15/2004	21 59.5	−39 23	Planetary nebula; a cross between the Ring Nebula and the Helix Nebula
L290	NGC 1261	07/15/2004	03 12.3	−55 13	Globular cluster; makes a pattern with Achernar and the Magallanic clouds
*L291	NGC 1493	07/15/2004	03 57.5	−46 12	Very comet-like galaxy
L292	NGC 2903	09/16/2004	09 32.2	+21 30	Elongated galaxy; long and dusty—Judith Irwin at Queen's University is doing a major radio study on this galaxy, involving the most sensitive HI observations of any galaxy; this galaxy was chosen because it takes up a fairly large amount of sky, but at the same time is far enough not be a part of the local group[1]

L293	NGC 6637 M69	10/07/2004	18 31.4	−32 21	Spotted by Wendee on 10/15/2004; Globular cluster
L294	NGC 6652	10/07/2004	18 35.8	−32 59	Both this and M69 are just SE of "bright" stars and are close to each other
L295	NGC 6726	10/07/2004	19 01.7	−36 53	Emission/reflection nebula in two parts; beautiful object close to globular cluster NGC 6723, L94
L296	NGC 2362	10/07/2004	07 18.8	−24 57	Open Cluster, faint but with bright, probably foreground star. This cluster seems Y-shaped
L297	47 Ursae Majoris	05/01/1964	10 59.7	+40 24	A Sun-like star that has at least two planets orbiting it in almost circular orbits
L298	NGC 6584	10/13/2004	18 18.6	−52 13	Globular cluster
L299	NGC 6273 M19	10/13/2004	17 02.6	−26 16	Globular cluster
L300	NGC 3226	11/08/2004	10 23.4	+19 54	Round galaxy—close companion; really neat field; striking view

> Last night of all,
> When yond same star that's westward from the pole
> Had made his course that part of heaven
> Where it now burns, Marcellus and myself,
> The bell then beating one.
>
> —Shakespeare, *Hamlet* 1.1.35–39, circa 1600,
> possibly alluding to the supernova of 1572
> which was "westward from the pole" on a
> November night in 1572, when Shakespeare was eight

> A star, a daystar, a firedrake rose at his birth. It shone by day in the heavens alone, brighter than Venus in the night, and by night it shone over delta in Cassiopeia, the recumbant constellation which is the signature of [Shakespeare's] initial among the stars. His eyes watched it, lowlying on the horizon, eastward of the bear, [but westward from Polaris] as he walked by the slumberous summer fields at midnight.
>
> —Joyce, *Ulysses*, 1922, writing about
> Shakespeare's possible reaction to the 1572 supernova

L301	**NGC 2670**	11/08/2004	08 45.5	−48 47	Open cluster; looks like a bow and arrow!
L302	**NGC 3114**	11/13/2004	10 02.7	−60 07	Open cluster; bright but scattered (M=4.2; D=35′)
L303	**NGC 4349**	11/13/2004	12 24.5	−61 54	Open star cluster; big, bright, and scattered
L304	**NGC 4103**	11/13/2004	12 06.7	−61 15	Open star cluster; bright and scattered
L305	**NGC 4755**	11/13/2004	12 53.6	−60 20	The Jewel Box open cluster; one of youngest—age 7.1 million years; gorgeous; dense; few bright stars in the form of letter A with faint stars mostly on east side of the A
L306	**NGC 2808**	11/13/2004	09 12.0	−64 52	Globular cluster; highly resolved
L307	**NGC 2516**	11/17/2004	07 58.3	−60 52	Southern Beehive, at tip of the "false cross" asterism; absolutely stunning view— Carolyn Shoemaker's favorite this night
L308	**NGC 3324**	11/17/2004	10 37.3	−58 38	Open cluster with nebulosity
L309	**NGC 1316**	11/18/2004	03 22.7	−37 12	A big elliptical galaxy that is devouring a spiral, as indicated by the dust lanes typical of a spiral; a number of small globular clusters could be from an even older galaxy that 1316 cannibalized long ago; it is the brightest member of the Fornax Cluster
L310	**NGC 2070**	11/18/2004	05 38.6	−69 05	Tarantula Nebula and cluster, in the Large Magellanic Cloud
L311	**IC2714**	11/18/2004	11 17.9	−62 42	Open cluster; large and irregular; close to Melotte 105 (below)
L312	**Melotte 105**	11/13–18/2004	11 19.4	−63 29	Open cluster; near IC 2714; faint, broad, one star in field to N; an irregular unresolved patch of light at low power; one star in field to north; took two nights to figure out which was which with these two

*L313	**IC4499**	11/18/2004	15 00.3	−82 13	Globular cluster in Apus; south-ernmost globular cluster
L314	**NGC 7213**	22:09:18	−47	10	Seyfert galaxy in Grus
L315	**NGC 5146**	12/15/2004	13 26.5	−12 19	Round galaxy
L316	**NGC 1976**	09/09/1962 12/17/2004	05 35.3	−05 23	Great Nebula in Orion
L317	**NGC 5838**	12/17/2004	15 05.4	+02 06	Very elongated galaxy with bright core
L318	**NGC 2841**	12/17/2004	09 22.0	+50 58	Elongated galaxy
L319P	**ESO573-12**	12/17/2004	12 20.6	−18 40	Small galaxy close to TV Corvi; see L73
L320	**Tombaugh 1**	01/08/2005	07 00.5	−20 34	Open cluster (distance: 4,100 ly); 45 stars
L321	**Tombaugh 2**	01/08/2005	07 03.1	−20 49	Open cluster (distance: 43,000 ly); 50 stars
L322	**IC166 Tombaugh 3**		01/05/2005	01 52.5	+61 50 Open cluster
L323	**Tombaugh 4**	01/08/2005	02 29.2	+61 47	Open cluster Nebula complex IC1795 just a little to west
L324	**Tombaugh 5**	01/05/2005	03 47.8	+59 03	Open cluster
L325	**IC1795**	01/08/2005	02 26.5	+62 04	Difficult to see visually but remarkable photographic target
L326	**NGC 4570**	01/08/2005	12 36.9	+07 15	On page 130 of *Starlight Nights*, Leslie Peltier recommends this object as something that could be confused with a comet
L327P	**IC2531**	01/08/2005	09 59.9	−29 37	Lenticular galaxy
L328P	**ESO435-16**	01/08/2005	09 58.6	−28 37	Peculiar galaxy; part of the NGC 2997 galaxy group; an unusual flare or arm, almost as large as the galaxy's central hub, spreads out toward the west, but no corresponding feature to the east; or ESO 095632-2822.8
L329	**T CrB**	01/08/2005	15 39.5	+25 55	Recurring nova in Corona Borealis
L330	**NGC 457**	01/10/2005	01 19.1	+58 20	ET, Dragonfly, or Phi Cas open cluster
L331P	**NGC 3319**	01/17/2005	10 39.2	+41 41	In CN1 area 377; graceful spiral galaxy with HII regions and sweeping arms; gorgeous

L332P West lobe of "Tombaugh's Great Stratum" near Alpheratz

	UGC371	01/17/2005	00 37 21.2	+29 09	A faint, nondescript edge-on spiral, which happens to be in the midst of the "Pegasus-Andromeda lobe" of Clyde Tombaugh's "stratum" of galaxies

L333P East lobe of "Tombaugh's Great Stratum" near Algol
PGC2218765 and PGC2218178

	01/17/2005	03 14.4	+43 21	Round galaxy
		03 14.4	+43 20	Perseus spiral galaxy; these faint galaxies are in the midst of the "Perseus lobe" of Clyde Tombaugh's "stratum" of galaxies; PGC is the Palomar Green Catalog
*L334 NGC 6342	01/17/2005	17 21.2	−19 35	SSE of Messier 9; weak globular; Shapley class 4
L335 Gegenschein	08/20/1966			variable
L336 M45	09/01/1961	03 47.0	+24 07	Pleiades Cluster with reflection nebulosity
L337 Q0957+56A/B	02/08/2005	10 01.3	+55 54	Double quasar in Ursa Major
L338 NGC 1042	07/03/2005	02 40.4	−08 26	Low surface brightness galaxy in Cetus; involved with nearby NGC 1035
L339 NGC 891	07/03/2005	02 22.6	+42 21	Extremely elongated galaxy in Andromeda
L340 NGC 1134	07/05/2005	02 53.6	+13 00	Aries round galaxy; low surface brightness
L341 NGC 755	07/08/2005	01 56.4	−09 04	Cetus elongated galaxy; found at low power despite being at magnitude 12.6
L342P NGC 91	07/14/2005	00 21.8	+22 25	Galaxy in Andromeda; part of gorgeous cluster of galaxies
L343 NGC 6438	08/28/2005	18 26.0	−85 25	Round galaxy in Octans; near the pole
L344 NGC 1313	08/28/2005	03 18.3	−66 30	Beautiful barred spiral in Reticulum, in which a burst of star formation is taking place; striking even through a 6-inch telescope
L345 NGC 2031	08/28/2005	0533.7	−70 59	Mensa open cluster with nebulosity in the Large Magellanic Cloud (LMC)
L346 NGC 2103	08/28/2005	05 41.6	−71 20	Fainter Mensa open cluster in LMC
L347 NGC 2060	08/28/2005	05 37.6	−69 10	Open cluster with nebulosity and dust; associated with Tarantula Nebula in LMC
L348 NGC 1866	08/28/2005	05 13.5	−65 28	Open cluster in Dorado, in LMC
L349 NGC 2547	08/28/2005	08 10.7	−49 16	"Bow and Arrow" open cluster in Vela; distance 2,000 light-years; sharply resembles a bow and arrow!

L350	NGC 4372	08/28/2005	12 25.8	−72 40	Globular cluster in Musca; distance 19,000 light-years; loosely concentrated
L351	NGC 6752	08/28/2005	19 10.9	−59 59	The great cluster in Pavo; distance 13,000 light-years
L352	NGC 6397	08/28/2005	17 40.7	−53 40	The great cluster in Ara; very large globular because it is only 7,200 light-years away; it is only slightly farther away than M4, the closest; stretches beautifully through high-power field of view of 6-inch f/4; (these last two were introduced to me by Lance Humphreys and later picked up during a comet sweep)
L353	Wendee's fishhook	08/29/2005	11:36.6	−63 02	Joined to NGC 3766 (see Levy 101)

Asterism with IC 2944 and IC2948

by a string of stars consisting of HIP (for Hipparcos) 56556, NGC 3766, HIP 56754, 56986, 57175, 57211, 57108, 56897, IC2948, IC2944, HIP 56726, and Lambda Centauri; Wendee found this string in April 2005 and described it as a "reversed J" or "fishhook"; through binoculars, loops of dark nebulae stretching out from Eta Carinae reach this asterism. NGC 3766 is at the eye of the hook; the two IC nebulae are at the bend near the barb, and the bright star Lambda Centauri is at the point; IC 2948 is a rich star-forming region with a plethora of Bok globules

*L354	NGC 5927	08/29/2005	15 28.0	−50 40	Globular cluster; resembles a faint comet in 6-inch f/4; distance 25,000 light-years
L355	Proxima Cen	08/29/2005	14 30.2	−62 42	The nearest star to the Sun; it is near an isosceles triangle of stars (TYC 9010–1420–1, 9010–1732–1, and 9010–1860–1); draw a line from 1420 to 1860 (the base of the triangle) and extend it about four times until you reach a reddish 10.8 magni-

					tude star in the midst of a field of Milky Way stars
L356	Barnard 263	08/29/2005	17 26.9	−42 37	A stunningly black nebula; optically very thick and blocking out virtually all the stars behind it; noticeable even in 6-inch f/4 at low power
L357	NGC 6281	08/29/2005	17 04.8	−37 54	"A-frame" open cluster (with nebulosity) in Scorpius
L358	NGC 6231	08/29/2005	16 54.0	−41 48	The "Big Arch" open cluster with nebulosity in Scorpius; faintly visible to the naked eye
L359	NGC 1566	08/29/2005	04 20.0	−54 56	Spiral galaxy in Dorado; distance about 44 million light-years; beautifully symmetrical spiral arms
L360	NGC 6684	08/30/2005	18 49.0	−65 11	Round galaxy in Pavo
L361	NGC 1763	08/30/2005	04 56.8	−66 24	Dorado nebula with dust and cluster, in LMC
L362	NGC 1734	08/30/2005	04 53.3	−68 47	Dorado open cluster in LMC
L363	NGC 1433	08/30/2005	03 42.0	−47 13	Horologium elongated galaxy
*L364	NGC 1527	08/30/2005	04 08.4	−47 53	Horologium very elongated galaxy with bright core; nice to see this in 6-inch f/4
*L365	NGC 1512	08/30/2005	04 03.9	−43 21	Horologium round galaxy
*L366	NGC 1808	08/30/2005	05 07.7	−37 31	Columba very elongated galaxy; brighter core
L367	NGC 613	08/31/2005	01 34.3	−29 25	Sculptor spiral galaxy; includes some bright knots
*L368	NGC 1448	08/31/2005	03 44.5	−44 39	Horologium edge-on galaxy; appeared at the very edge of the field; when I brought the galaxy to the center of the 6-inch, it was a lovely, faint string of fuzzy light.
L369	IC2177	08/31/2005	07 05.1	−10 42	Reflection nebula in Monoceros; long and beautiful, cutting across the field beneath an open cluster like stranded rope
*L370	NGC 5643	09/01/2005	14 32.7	−44 10	Lupus round galaxy
L371	NGC 1617	09/01/2005	04 31.7	−54 36	Dorado elongated galaxy
L372	NGC 2025	09/01/2005	05 33.1	−71 44	Mensa Open cluster in LMC
L373	NGC 6362	09/02/2005	17 31.9	−67 03	Ara globular cluster; distance 17,000 light-years; easy to resolve
L374	NGC 6744	09/02/2005	19 09.8	−63 51	Pavo round galaxy

L375	**NGC 5189**	09/03/2005	13 33.5	−65 59	Oddly shaped planetary nebula in Musca; through 6-inch f/4 looked like a star cluster in nebulosity, but not stars; nighly irregular shape in small telescope, spiral in larger instruments; distance about 3,000 light-years
L376	**X TrA**	09/03/2005	15 14.3	−70 05	Strikingly red carbon star; range 5.0–6.4; irregular
*L377	**NGC 6300**	09/03/2005	17 17.0	−62 49	Ara low surface brightness galaxy; large, like a faint comet in small telescope
*L378	**IC5267**	09/03/2005	22 57.2	−43 24	Round galaxy; bright core
L379	**NGC663**	09/16/1962	01 46.0	+61 15	The Horseshoe Open Cluster in Cassiopeia
L380	**NGC6715 M54**	07/29/1965	18 55.1	−30 29	Globular cluster in Sagittarius; in 1994 Ibata, Gilmore, and Irwin (*Nature* 370, 21 July 1994) discovered what is now known as the Sagittarius Dwarf Elliptical Galaxy; the Milky Way is cannibalizing this galaxy; in 1995 the team found that M54, discovered by Messier in 1778, is a gigantic globular cluster within one "knot" of this galaxy, and is 88,000 light years away
L381	**UGC5470**	12/01/1984	10 08.5	+12 18	Leo I or Regulus Galaxy, very difficult, low surface brightness, near Regulus; dE3 galaxy in Local Group; Leo I and II discovered by Harrington and Wilson on the Palomar Sky Survey in 1950
L382	**UGC6253**	12/01/1984	11 13.5	+22 10	Leo II galaxy; dE0 peculiar galaxy in Local Group; also very difficult visually in 16-inch
L383	**NGC1049**	09/22/2005	02 39.7	−34 17	Globular cluster in the Fornax System, a dwarf galaxy; discovered by John Herschel in the 1830s during his time at the Cape of Good Hope; Harlow Shapley found the galaxy of which it is a part in 1938; cluster is about 500,000 light-years away; (M = 12.6, D = 0.7');I spotted it with both Miranda, the 16-inch f/5, and Minerva, the 6-inch f/4

The list which follows arranges the entire list in order of position in the sky by right ascension, which will allow the reader to plan an observing session using the accompanying atlas.

LEVY'S CATALOG OF COMET MASQUERADERS AND MORE
(with thanks to Bill Logan)

L (for Levy) = David Levy's observation numbers
NGC = New General Catalog
IC = Index Catalog
UGC = Uppsala General Catalog
ESO = European Southern Observatory Catalog

NGC	Messier	Levy	Other	RA	DEC	Mag	Dia	Chart
		L335	Gegenschein	varies	varies			
7814		*L276		00 03.3	+16 09	10.6	5.3'	1
0091		L342P		00 21.8	+22 25	15~	2.5'	
0104		L104		00 24.1	−72 05	4.0	47.0'	29
0147		*L081		00 33.2	+48 30	10.4	14.0'	
0150		*L223		00 34.3	−27 48	11.4	3.8'	3, 4, 24
0157		*L240		00 34.8	−08 24	10.4	4.1'	3, 24
		L332P	UGC371	00 37.3	+29 09	15.0	?	1
0185		*L080		00 39.0	+48 20	9.2	12.0'	2
0224	M31	L086		00 42.7	+41 16	4.3	180'	1, 2
0246		L061		00 47.0	−11 53	8.5	240'	3, 4
0247		*L239		00 47.1	−20 46	9.1	21.0'	3, 4, 24

0253		L092		00 47.6	−25 17	8.0	6.0'	3, 4, 24
0262		L272	Markarian 348 Swift II	00 48.8	+31 57	15p	1.4	1
		L252	65 Piscium	00 49.9	+27 43	6.3	6.3'	1
0270		L056		00 50.6	−08 39	12.9	1.8'	3
0289		*L222		00 52.7	−31 12	11.0	5.0'	3, 4, 24
0281		L194	IC1590	00 52.8	+56 37	7.0	35'	1, 2, 13
0300		*L221		00 54.9	−37 41	8.1	12.0'	4
0362		L105		01 03.2	−70 51	6.4	13.0'	
0404		L064		01 09.4	+35 43	10.3	3.4'	1, 2
0457		L330		01 19.1	+58 20	6.4	13.0'	1, 2
0470		L278		01 20.1	+03 25	12.5	2.9'	1
0474		L278		01 20.3	+03 26	12.3	6.1'	
0488		L034		01 21.8	+05 15	10.3	5.2'	1, 3
0514		L031		01 24.1	+12 55	12.3	3.7'	1
0524		L059		01 24.8	+09 32	10.6	3.2'	1
0578		*L238		01 30.5	−22 40	10.9	4.7'	3, 4
0598	M33	L091		01 33.9	+30 39	5.7	60.0'	1
0613		L367		01 34.3	−29 25	10.1	6.3'	
0628	M74	L060		01 36.6	+15 47	9.8	10.0'	1
		L322	IC166	01 52.5	+61 50	11.7	7.0'	2, 13

0718				01 53.2	+04 12	11.7	2.2'	1, 3
0755		L341		01 56.4	−09 04	12.6	4.0'	
0752		L053		01 57.8	+37 41	5.7	50.0'	1, 2
		L241	AA Ceti	01 59.0	−22 55	?	n/a	3, 4
0772		*L058		01 59.3	+19 01	10.3	7.1'	1
0869				02 19.0	+57 09	4.4	30.0'	1, 2
0884		L254		02 19.0	+57 09	4.7	30.0'	1, 13
0891		L339		02 22.6	+42 21	10.5		
0898		L062		02 23.3	+41 57	14.4	1.0'	1, 2, 5
		L235	IC1795	02 26.5	+62 04	?	20.0'	1, 2, 12
0936		L109		02 27.6	−01 09	10.7	5.2'	3
		L323	Tombaugh 4	02 29.2	+61 47	?	5.3'	
0949		*L099		02 30.8	+37 08	12.4	2.7'	2, 5
		L224	IC1830	02 39.1	−27 27	12.8	1.7'	3, 4, 28
1023		L068		02 40.4	+39 04	10.5	8.7'	2, 5
1042		L338		02 40.4	−08 26			
1068	M77	L111		02 42.7	−00 01	8.8	6.9'	3
1097		L288		02 46.3	−30 17	9.3	9.3'	3, 4, 6, 28
1134		L340		02 53.6	+13 00			
1165		*L242		02 58.7	−32 06	12.7	2.3'	3, 4, 6, 28

1187		L225		03 02.6	−22 52	10.8	5.3'	3, 4, 6, 28
		L100	Beta Persei	03 08.1	+40 58	2.02		2, 5
1261		L290		03 12.3	−55 13	8.4	6.9'	28, 29
		L333P	PGC2218765	03 14.4	+43 21	15.5	0.2'	2, 5
		L333P	PGC2218178	03 14.4	+43 20	15.0	0.2'	
1313		L344		03 18.3	−66 30	8.7	9.0'	
1316		L309		03 22.7	−37 12	6.7	16.0'	4, 28
1333		L044		03 29.2	+31 25	?	9.0'	5, 7
1360		L286		03 33.3	−25 51	14.0	6.5'	2, 3, 6, 28
1365		L287		03 33.6	−36 08	9.5	9.8'	6, 28
1380		L200		03 36.5	−34 59	11.0	4.9'	4, 6, 28
1399		L200		03 38.5	−35 27	9.9	3.2'	4, 6, 28
1404		L200		03 38.9	−35 35	10.3	2.5'	4, 6, 28
		L283	U Camelopardalis	03 41.7	+62 40			1, 2, 13
1433		L363		03 42.0	−47 13	9.8	6.3'	
1448		*L368		03 44.5	−44 39	10.7	7.4'	
	M45	L336	Pleiades	03 47.0	+24 07			7
		L324	Tombaugh 5	03 47.8	+59 03	8.4	15.0'	2, 13
1493		*L291		03 57.5	−46 12	11.8	3.6'	4, 28, 29
1491		L049		04 03.4	+51 19	?	3.0'	1, 2, 5

1512		*L365		04 03.9	−43 21	10.3	8.5'	
1527		L364		04 08.4	−47 53	10.8	3.5'	
1566		L359		04 20.0	−54 56	9.7	8.2'	
1579		L050		04 30.2	+35 16	?	3.0'	5, 13
1600		L123		04 31.7	−05 05	12.0	3.1'	6, 8
1617		L371		04 31.7	−54 36	10.4	4.2'	
1624		L008		04 40.4	+50 27	11.8	1.9'	5
1637		*L114		04 41.5	−02 51	10.9	3.9'	6, 8
1679		L047		04 50.0	−31 59	12.0	2.7'	6, 28
1734		L362		04 53.3	−68 47		12'	
1763		L361		04 56.8	−66 24			
		L148	R Leporis	04 59.6	−14 48			6, 28
1746		L273		05 03.6	+23 49	6.0	42'	5, 7
1788		*L115		05 06.9	−03 21	11.0	8.0'	6, 8
1808		L366		05 07.7	−37 31	9.9	6.3'	
1866		L348		05 13.5	−65 28	9.8	4.7'	
1851		L046		05 14.0	−40 03	7.1	11.0'	28
1904	M79	L025		05 24.5	−24 33	7.7	7.8'	6, 28
1931		L001		05 31.4	+34 15	9.5	1.0'	5, 7, 10
2025		L372		05 33.1	−71 44	10.9	1.8'	

2031		L345		05 33.7	−70 59	10.8	3.3'	
1952	M1	L033		05 34.5	+22 01	8.4	8.0'	5, 7, 10
1976	M42	L316		05 35.3	−05 23	4.0	90.0'	6, 8
1999		L095		05 36.5	−06 42	10.0	16.0'	6, 8
2060		L347		05 37.6	−69 10			
2070		L310		05 38.6	−69 05	5.0	40.0'	29
2023		L125	IC434	05 41.6	−02 14	?	10.0'	6, 8
2103		L346		05 41.6	−71 20			
2068	M78	*L014		05 46.7	+00 03	8.0	8.0'	8
2158		L116		06 07.5	+24 06	11.0	5.0'	5, 7, 10
2174		L124		06 09.7	+20 30	?	40.0'	5, 7, 10
2188		*L045		06 10.0	−34 06	12.1	4.7'	6, 9, 28
		L113	TU Geminorum	06 10.9	+26 01	9.4		5, 7, 10
2237		L077		06 32.3	+05 03	?	?	8
2245		*L161		06 32.7	+10 10	?	2.0'	8
2252		L162		06 35.0	+05 23	7.7	20.0'	8
2254		L160		06 36.0	+07 40	9.1	4.0'	8
2261		*L078		06 39.2	+08 44	?	2.0'	8
2264		L159		06 41.1	+09 53	3.9	59'	
		L320	Tombaugh 1	07 00.5	−20 34	?	6.0'	9

		L321	Tombaugh 2	07 03.1	−20 49	?	3.0'	9
		L369	IC2177	07 05.1	−10 42		110'	
2359		L126		07 18.6	−13 12	?	10.0'	
2362		L296		07 18.8	−24 57	10.5	8.0'	9
		L018		07 29.2	+20 55	9.3	47.0'	10
2392		*L158	IC2194	07 33.7	+31 19	15.0	0.9'	10
		*L158	IC2196	07 34.1	+31 24	14.0	1.4'	10
		*L158	IC2197	07 34.3	+31 24	14.0	0.4'	10
2403		L009		07 36.9	+65 36	8.9	18.0'	13
2419		L195		07 38.1	+38 53	10.3	4.1'	10
2420		L035		07 38.5	+21 34	10.2	10.0'	10, 13
2437	M46	L198		07 41.8	−14 49	6.1	27'	9
2438		L198		07 41.8	−14 44	10.8	1.1'	
		L227	Haffner 17	07 51.6	−31 49			9, 25
2477		L258		07 52.3	−38 33	5.8	26'	25
2516		L307		07 58.3	−60 52	3.8	30.0'	25
2547		L349		08 10.7	−49 16	4.7	19'	
		L256	IC2367	08 24.2	−18 46	12.5	2.3'	15
2613		L226		08 33.4	−22 58	10.3	7.1'	15, 25
2670		L301		08 45.5	−48 47	7.8	9.0'	25

2683		*L093		08 52.7	+33 25	9.8	9.0'	11
2681		L096		08 53.5	+51 19	11.1	3.7'	13
2655		L010		08 55.6	+78 13	11.0	4.9'	13
2775		*L163		09 10.3	+07 02	10.1	4.2'	11, 15
2808		L306		09 12.0	−64 52	6.3	13.8'	26, 29
2841		L318		09 22.0	+50 58	9.3	8.1'	12, 13
2903		L292		09 32.2	+21 30	9.6	12.0'	11
		L051	Arp 321	09 38.9	−04 52	13.8	2.6'	15
2964		*L131		09 42.9	+31 51	11.3	2.7'	11
2968		*L131		09 42.9	+31 51	11.3	2.7'	
2986		*L169		09 44.3	−21 17	10.8	3.1'	15, 25
		L248	Delta Scorpii	16 00.5	−22 38	2.28		20
3049		*L209		09 54.8	+09 16	13.3	2.2'	11, 15
3055		*L037		09 55.3	+04 16	12.6	2.1'	15
3031	M81	L024		09 55.6	+69 04	7.8	11.0'	12, 13
3034	M82	L022		09 55.8	+69 41	9.2	10.0'	12
3070		*L133		09 58.0	+10 22	12.3	1.3'	11, 15
		L328	ESO435-16	09 58.6	−28 37	13.3	2.0'	25
		L327	IC2531	09 59.9	−29 37	12.5	5.0'	25
		L129	UGC5373	10 00.0	+05 20	12.0	?	11, 15

3114		L302		10 02.7	−60 07	4.2	35'	25, 26, 29
3115		L052		10 05.2	−07 43	10.1	7.3'	15
3201		L218		10 17.6	−46 25	6.8	18.2'	25, 26
3198		*L130		10 19.9	+45 33	10.3	8'	12, 13
3226		L300		10 23.4	+19 54	12.4	2.5'	11
3242		*L168		10 24.8	−18 38	7.7	4.0'	15
3245		L165		10 27.3	+28 30	10.8	3.1'	11
3309		L219		10 36.6	−27 31	12.6	2.4'	14, 25
3311		L219		10 36.7	−27 32	12.6	3.7'	14, 25
3312		L219		10 37.0	−27 34	12.7	3.3	14, 25
3324		L308		10 37.3	−58 38	6.3	13.8'	25, 26, 29
3314		L219		10 37.4	−27 41	13.5	1.5'	14, 25
3316		L219		10 37.6	−27 36	13.7	1.4'	14, 25
3310		L167		10 38.7	+53 30	10.8	4.2'	12, 13
3319		L331P		10 39.2	+41 41	11.5	6.1'	11, 12, 13
		L102	IC2602	10 43.2	−64 24	1.9	102'	26, 29
3344		*L166		10 43.5	+24 55	10.5	7.1'	11
3372		L075		10 45.1	−59 41	3.0	120'	25, 26, 29
		L072	V Hydrae	10 51.6	−21 15	?	?	14, 15, 25
3432		L132		10 52.5	+36 37	11.3	6.6'	11, 12

		L297	47 Ursae Majoris	10 59.7	+40 24	5.0	?	11, 12, 13
		L337	Q0957+56A/B	10.01. 3	+55 54			5, 7, 12, 13
3486		*L164		11 00.4	+28 58	10.5	6.9'	11
3556	M108	L089		11 11.5	+55 40	10.6	8.6'	12, 13
3587	M97	L088		11 14.8	+55 01	12.0	3.2'	12, 13
		L311	IC2714	11 17.9	−62 42	8.0	12.0'	26, 29
3621		L103		11 18.3	−32 49	10.1	12.0'	14, 25
3623	M65	L074		11 18.9	+13 05	10.1	9.0'	11, 15
		L312	Melotte 105	11 19.4	−63 29	8.5	4.0'	26, 29
3627	M66	L016		11 20.2	+12 59	9.7	9.1'	11
3628		L079		11 20.3	+13 36	10.4	13.0'	11, 15
3718		*L137		11 32.6	+53 04	10.8	7.9'	13
3738		*L136		11 35.8	+54 31	11.7	2.3'	12, 13
3766		L101		11 36.1	−61 37	5.0	14.0'	26
		L353	"The Fish Hook"	11 36.6	−63 02			
3810		L041		11 41.0	+11 28	11.3	4.1'	
3865		*L201		11 44.9	−09 14	13.0	2.3'	14
3887		*L181		11 47.1	−16 51	10.6	3.1'	14
3923		L217		11 51.0	−28 48	9.8	5.8'	14
3953		*L138		11 53.8	+52 20	10.6	6.9'	12, 13

3992	M109	L090		11 57.6	+53 23	10.6	7.5'	12
4038		L106		12 01.9	−19 52	10.9	3.4'	
4039		L106		12 01.9	−19 52	11.0	3.3'	
4103		L304		12 06.7	−61 15	7.0	7.0'	26, 29
4129		*L193		12 08.9	−09 02	12.5	2.2'	14
		L108	Struve 1604	12 09.5	−11 52	?	?	14
4214		L234		12 15.6	+36 20	9.7	7.9'	12
4244		L235		12 17.5	+37 49	10.2	16.2'	
4256		L135		12 18.7	+65 54	11.9	4.2'	12
4258	M106	L026		12 19.0	+47 18	9.1	17.0'	12
4274		*L142		12 19.8	+29 37	10.3	6.6'	16
		L073	*TV Corvi	12 20.4	−18 27	?	?	14
		L319	ESO573-12	12 20.6	−18 40	?	?	
4319		L134		12 21.7	+75 19	12.8	2.8'	13
4303	M61	*L180		12 21.9	+04 28	9.7	6.3'	17
4340		L042		12 23.6	+16 43	12.1	3.1'	16, 17
4361		L107		12 24.5	−18 48	10.3	45.0'	14
4349		L303		12 24.5	−61 54	7.4	16'	26, 29
4374	M84	L065		12 25.1	+12 53	10.2	6.7'	
		L229	SS Virginis	12 25.2	+00 46	7.68	?	14

4372		L350		12 25.8	−72 40	7.8	18'	
4406	M86	L066		12 26.2	+12 57	9.9	9.8'	16, 17
4414		L233		12 26.4	+31 13	10.3	3.6'	16
4449		L139		12 28.2	+44 06	9.6	6.1'	12
4450		L171		12 28.5	+17 05	10.1	5.0'	16, 17
		L230	3C273 Virginis	12 29.1	+02 03	?	?	17
4473		*L145		12 29.8	+13 26	10.2	4.2'	16, 17
4472	M49	*L146		12 29.8	+08 00	8.4	8.9'	17
4485		L140		12 30.5	+41 42	11.9	2.2'	
4490		*L228		12 30.6	+41 38	9.8	6.1'	12
4486	M87	L175		12 30.8	+12 24	8.6	6.9'	16, 17
4501	M88	*L144		12 32.0	+14 25	9.5	5.8'	16, 17
4519		L260		12 33.5	+09 39	11.7	3.1'	16, 17
4535		*L179		12 34.3	+08 12	10.0	6.9'	
4536		*L192		12 34.5	+02 11	10.6	7.4'	
4598	M91	*L173		12 35.4	+14 30	10.2	4.2'	
4552	M89	L177		12 35.7	+12 33	9.8	5.0'	
4559		*L143		12 36.0	+27 58	10.0	10.5'	16
4565		L141		12 36.3	+25 59	9.6	16.2'	16, 17
4567		L199		12 36.5	+11 15	11.3	3.0'	16, 17

4568		L199		12 36.5	+11 15	10.8	46'	17
4570		L326		12 36.9	+07 15	10.9	4.1'	17
4579	M58	*L176		12 37.7	+11 49	9.8	4.7'	16, 17
4591		*L188		12 39.3	+06 01	13.0	1.5'	17
4590	M68	L216		12 39.5	−26 45	8.2	12.0'	14
4596		*L178		12 39.9	+10 11	10.4	3.9'	16, 17
4605		L011		12 40.0	+61 37	10.8	5.9'	12, 13
4594	M104	L067		12 40.0	−11 37	8.2	6.9'	14
4623		*L213		12 42.2	+07 41	12.2	2.1'	17
4636		L182		12 42.8	+02 41	10.4	5.9'	17
4651		L170		12 43.7	+16 24	10.8	3.9'	16, 17
4649	M60	*L174		12 43.7	+11 33	8.8	7.1'	16, 17
4656		L063		12 44.0	+32 10	11.2	18.0'	16
4685		*L210		12 47.1	+19 28	12.6	1.5'	16, 17
4689		*L172		12 47.8	+13 46	10.9	4.2'	16, 17
4688		*L215		12 47.8	+04 20	11.9	3.1'	17
4713		*L214		12 50.0	+05 19	11.7	2.6'	17
4722		L259		12 51.5	−13 19	11.7	1.7'	14
4772		L191		12 53.5	+20 10	11.0	3.3'	16, 17
4755		L305		12 53.6	−60 20	4.2	9.5'	26, 29

4779		*L211		12 53.8	+09 44	12.4	2.0'	16, 17
4795		*L212		12 55.0	+08 04	12.1	1.8'	17
4826	M64	L015		12 56.7	+21 41	9.3	10.0'	16, 17
4818		*L183		12 56.8	−08 31	11.1	4.2'	14
4866		*L257		12 59.5	+14 10	11.2	6.1'	16, 17
4956		L190		13 05.1	+35 11	12.4	1.4'	12, 13
5016		L255		13 12.1	+24 06	12.8	1.6'	16, 17
5020		*L187		13 12.6	+12 36	11.7	3.1'	16, 17
5024	M53	L023		13 12.9	+18 10	7.6	22.0'	16, 17
5055	M63	L007		13 15.8	+42 02	9.3	13.0'	12
5068		*L232		13 18.9	−21 02	10.0	7.1'	14
5127		L189		13 23.8	+31 34	11.9	2.7'	12, 16
5128		L054		13 25.5	−43 01	7.6	28.0'	
5147		*L184		13 26.3	+02 06	11.8	1.8'	17
5146		L315		13 26.5	−12 19	12.3	1.7'	14
5139		L097		13 26.8	−47 29	3.7	45.0'	26
5194	M51	L087		13 29.9	+47 12	8.9	11.0'	12, 18
5189		L375		13 33.5	−65 59	9.9	2.6'	
5248		*L185		13 37.5	+08 53	10.3	6.1'	13, 17
5247		*L267		13 38.1	−17 53	10.0	5.0'	16, 17

5350		L261		13 53.4	+40 22	11.4	3.2'	12, 18
5371		*L186		13 55.7	+40 28	10.6	4.2'	
5364		*L231		13 56.2	+05 01	10.5	6.6'	17
5377		L027		13 56.3	+47 14	12.2	4.1'	3, 12, 18
5457	M101	*L002		14 03.2	+54 21	13.0	1.6'	12
5427		*L202		14 03.4	−06 02	11.4	2.5'	14
5473		*L028		14 04.7	+54 54	12.4	2.2'	12
5474		L029		14 05.0	+53 40	11.3	4.5'	
5466		*L040		14 05.5	+28 32	9.1	11'	18
5460		L269		14 07.6	−48 19	5.6	24'	
5634		L082		14 29.6	−05 59	9.5	8.4'	
5638		L083		14 29.7	+03 14	12.2	2.3'	
		L355	Proxima Centauri	14 30.2	−62 42	10.8		
5643		L370		14 32.7	−44 10	10.0	4.5'	
5676		L005		14 32.8	+49 28	11.9	3.9'	
5668		*L203		14 33.4	+04 27	11.5	3.3'	
5690		L268		14 37.7	+02 17	12.7	3.5'	
5694		L196		14 39.6	−26 32	10.2	4.3'	19
		*L313	IC4499	15 00.3	−82 13	10.6	7.4'	29
5838		L317		15 05.4	+02 06	10.8	4.2'	20

5846		L121		15 06.4	+01 36	13.8	30.0'	
5866	M102	L030		15 06.5	+55 46	15.2	2.7'	12
5850		*L204		15 07.1	+01 33	11.0	4.3'	20
		L376	X Tri Australis	15 14.3	−70 05	5.0–6.4		
5907		L197		15 15.9	+56 19	10.4	12.3'	12, 13
5897		L264		15 17.4	−21 01	8.6	12'	19, 20
5927		L354		15 28.0	−50 40	8.3	11'	
5962		L266		15 36.5	+16 37	11.3	2.9'	18, 20
		L329	T CrB	15 39.5	+25 55	9–10		18
5746		L265		15 44.9	+01 57	10.3	7.1'	20
6093	M80	L117		16 17.0	−22 59	7.3	13.0'	19, 20
6106		*L205		16 18.8	+07 25	12.2	2.6'	20
6118		*L206		16 21.8	−02 17	12.0	4.7'	20
6101		L285		16 25.8	−72 12	9.3	10.7'	29
6181		L236		16 32.3	+19 50	11.9	2.6'	18, 20
		L127	16/17 Draconis	16 36.2	+52 55	5.5		13
6207		L019		16 43.1	+36 50	12.1	3.0'	18
6229		L006		16 47.0	+47 32	9.4	5.4'	18
6231		L358		16 54.0	−41 48	2.6	230'	
6254	M10	L004		16 57.1	−04 06	6.6	21.0'	20

6273	M19	L299		17 02.6	−26 16	7.2	13.5'	19, 20, 23
6281		L357		17 04.8	−37 54	5.4	7.9'	
6287		L271		17 05.2	−22 42	9.3	5.0'	19, 20, 23
6300		L377		17 17.0	−62 49	10.2	4.2'	
6341	M92	L003		17 17.1	+43 08	6.4	15.0'	18
6342		*L334		17 21.2	−19 35	9.7	15.0'	19, 20, 23
6364		*L036		17 24.5	+29 24	14.1	1.5'	18
		L356	Barnard 263	17 26.9	−42 37			
6362		L373		17 31.9	−67 03	7.5	10'	
6384		*L207		17 32.4	+07 04	10.6	6.0'	20
6402	M14	L020		17 37.6	−03 15	7.6	33.0'	20
6397		L352		17 40.7	−53 40	5.6	26'	
6426		*L208		17 44.9	+03 00	11.2	3.2'	20
6440		L243		17 48.9	−20 22	9.7	5.4'	19
6445		L244		17 49.2	−20 01	13	0.6'	20, 23
6451		L120		17 50.7	−30 13	8.0	7.0'	
6514	M20	L152		18 02.3	−23 02	6.3	24'	19, 20, 23
		L118	Great Star Cloud	18 03.4	−27 54			19, 23
6522		*L279	Baade's Window	18 03.6	−30 02	8.6	16.0'	19, 23
6523	M8	L150		18 03.8	−24 23	5.0	76'	19, 20, 23

6535		L270		18 03.8	−00 18	10.6	3.5'	20
6528		*L279	Baade's Window	18 04.8	−30 03	9.5	17.0'	
6541		L274		18 08.0	−43 42	6.6	12'	19, 23
6553		L149		18 09.3	−25 54	8.3	7.9'	19, 20, 22, 23
	M24	L119		18 16.9	−18 29	11.5	1.3'	19, 20, 22, 23
6584		L298		18 18.6	−52 13	9.2	7.9'	23, 29
6611	M16	L151		18 18.8	−13 47	6.0	69'	19, 22, 23
6618	M17	L076		18 20.7	−16 10	7.0	25.0'	19, 20, 22, 23
6438		L343		18 26.0	−85 25	11.1	1.5'	
6638		L147		18 30.9	−25 30	9.0	6.6'	20, 22, 23
6637	M69	L293		18 31.4	−32 21	7.7	7.1'	19, 23
6652		L294		18 35.8	−32 59	8.6	4.5'	19, 23
6684		L360		18 49.0	−65 11	10.4	3.9'	
6709		L098		18 51.5	+10 21	6.7	13'	20, 22
6712		L084		18 53.1	−08 42	8.1	7.4'	20, 22
6720	M57	L013		18 53.6	+33 02	9.5	1.7'	21
6723		L094		18 59.6	−36 38	7.3	11'	
6726		L295		19 01.7	−36 53	?	78.0'	23
6744		L374		19 09.8	−63 51	8.3	18'	
6752		L351		19 10.9	−59 59	5.4	20'	

6760		L110		19 11.2	+01 02	9.1	6.3'	22
6781		L245		19 18.4	+06 33	11.4	1.8'	22
6791		L039		19 20.7	+37 51	9.5	15'	21
6814		L262		19 42.7	−10 19	11.2	3.2'	22, 23
6826		L122		19 44.8	+50 31	9.8	25.0'	13, 21
		L157	LWJ1948.2+3743	19 48.2	+37 43			21
6838	M71	L021		19 53.8	+18 47	8.2	9.0'	21
6853	M27	L017		19 59.6	+22 43	7.6	5.8'	21
6910		L220		20 23.1	+40 47	7.4	8.0'	21
		L128	IC5020	20 30.6	−33 29	13.0	2.9'	23, 24, 27
6934		L156		20 34.2	+07 24	8.9	5.8'	22
6946		L250		20 34.8	+60 09	8.9	11.0'	13
6992		L237	Veil Nebula	20 56.4	+31 43			21
6981	M72	L155		20 53.5	−12 32	9.4	5.8'	22, 24
7023		L112		21 00.5	+68 10	7.1	5.0'	2, 13
7006		L263		21 01.5	+16 11	10.6	2.8'	21, 22
7009		L154		21 04.2	−11 22	8.0	3.0'	20, 22, 24
		L070P	LWJ2108.8+0620	21 08.8	+06 20			22, 24
7078	M15	L012		21 30.0	+12 10	6.0	12.0'	22
		L085	IC1396	21 39.1	+57 30	?	12.0'	2

		L246	V460 Cygni	21 42.0	+35 31 5.7			21
		L289	IC5148	21 59.5	−39 23	13.0	2.0'	27
7184		L277		22 02.7	−20 49	10.8	5.8'	22, 24
		L069-P	LWJ2204.3+4508	22 04.3	+45 08			2, 21
7217		L247		22 07.9	+31 22	10.1	3.8'	1, 21
7213		L314		22:09.3	−47 10	10.5	1.9'	27
7293		L153		22 29.6	−20 48	7.3	860'	22, 24
7314		L282		22 35.8	−26 03	11.6	4.6'	24
7331		L032		22 37.1	+34 25	10.3	10.0'	1, 2
		L378	IC5267	22 57.2	−43 24	10.5	5.0'	
7457		L249		23 01.0	+30 09	11.2	4.2'	1
		L253	BC Andromedae	23 01.0	+46 31	?	?	2, 13
7582		L048		23 18.4	−42 22	11.3	5.0'	27, 29
7664		L055		23 26.6	+25 04	13.4	2.6'	4, 27, 29
7723		L057		23 38.8	−12 58	11.9	3.5'	3, 24
7721		*L280		23 38.8	−06 31	11.6	3.4'	3, 24
7727		L281		23 39.9	−12 18	11.5	4.7'	3, 24
		L071	LWJ2340.6+5618	23 40.6	+56 18			2, 13
		L275	TX Piscium	23 46.5	+03 29			3, 24
7753		L038		23 47.1	+29 20	13.0	2.9'	1

7789		L251		23 57.0	+56 44	6.7	15'	2, 12
7793		*L043		23 57.8	−32 35	9.7	9.6'	3, 24

When I heard the learn'd astronomer,
When the proofs, the figures, were ranged in columns before
    me . . .
How soon unaccountable I became tired and sick,
Till rising and gliding out I wander'd off by myself,
In the mystical moist night-air, and from time to time,
Look'd up in perfect silence at the stars.

                  —Walt Whitman,
         "When I Heard the Learn'd Astonomer," 1903

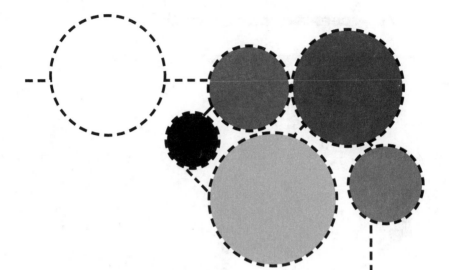

DEEP SKY
OBJECT ATLAS

309

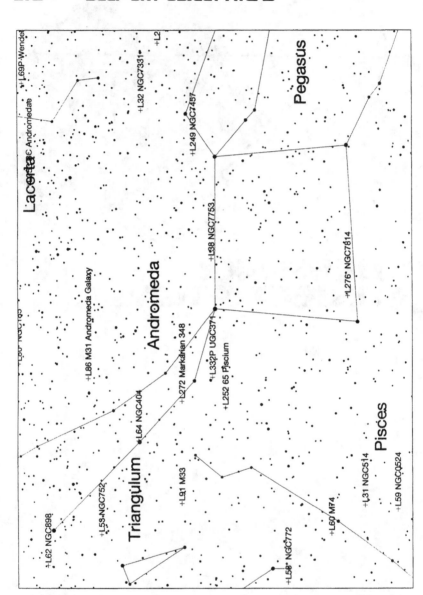

Chart 1: Pegasus, Andromeda, and northern Pisces

(The L is short for Levy in front of each object [e.g., L62 = Levy 62]. These charts were prepared by David and Wendee Levy, and we are grateful to Scott Roberts and Meade Instruments for permission to use their Autostar program for this purpose.)

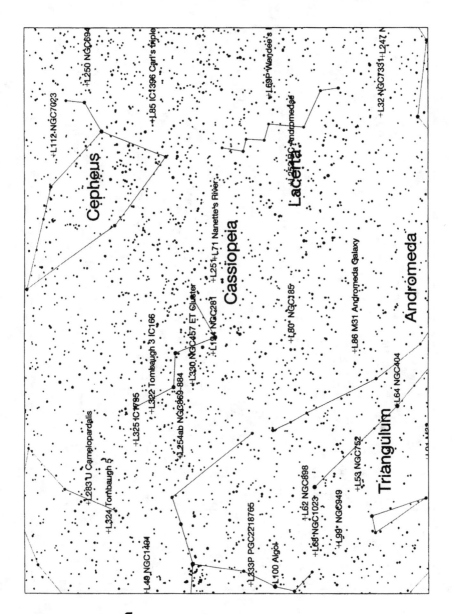

**Chart 2:
Cassiopeia,
Cepheus,
Camelo-
pardalis,
and Lacerta**

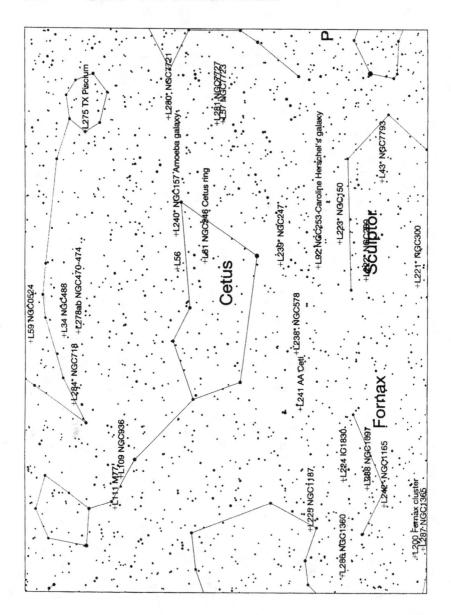

Chart 3:
Cetus

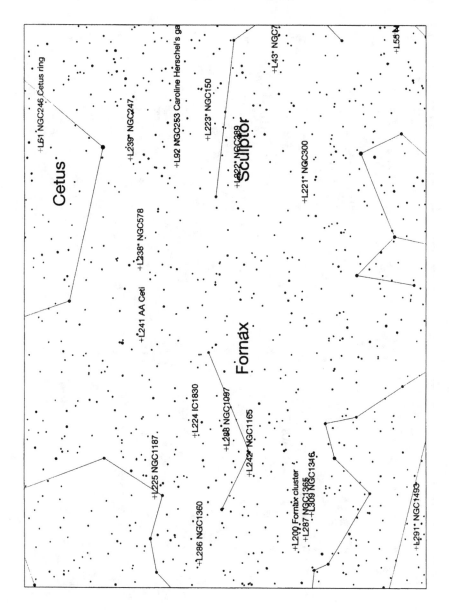

Chart 4:
Fornax and
Sculptor

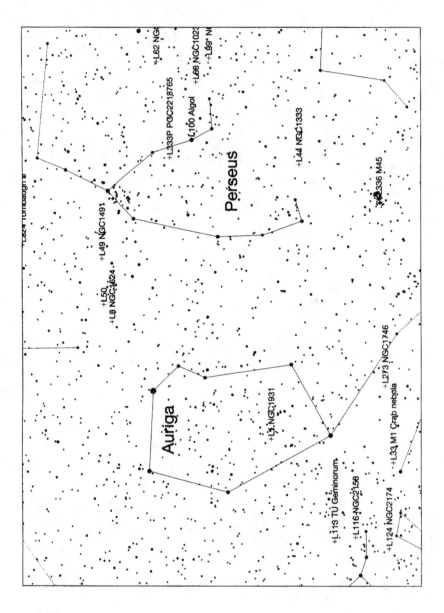

Chart 5:
Auriga and
Perseus

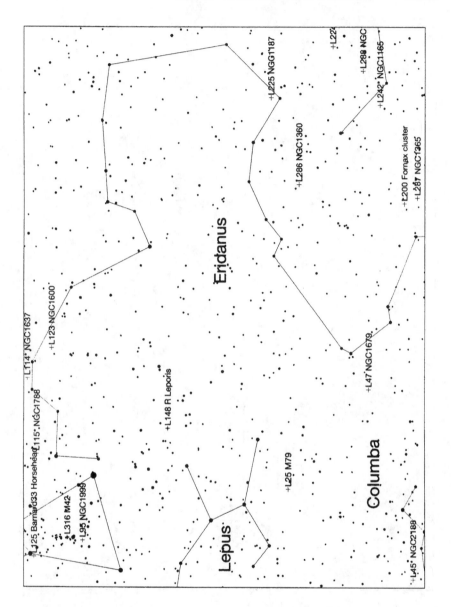

**Chart 6:
Lepus and
northern
Eridanus**

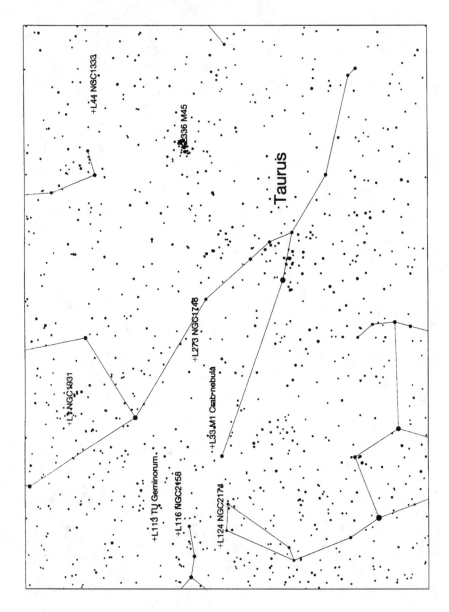

Chart 7:
Taurus

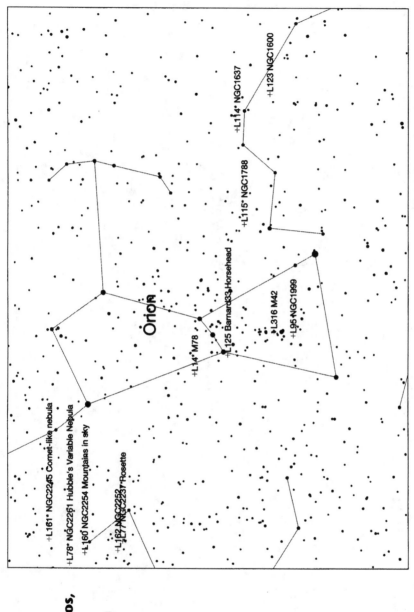

**Chart 8:
Orion,
western
Monoceros,
and
northern
Eridanus**

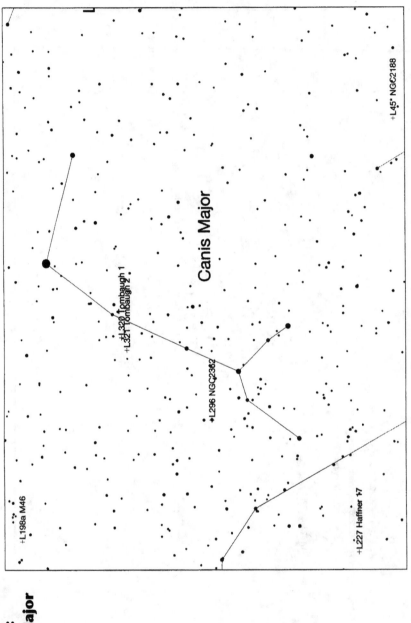

Chart 9:
Canis Major

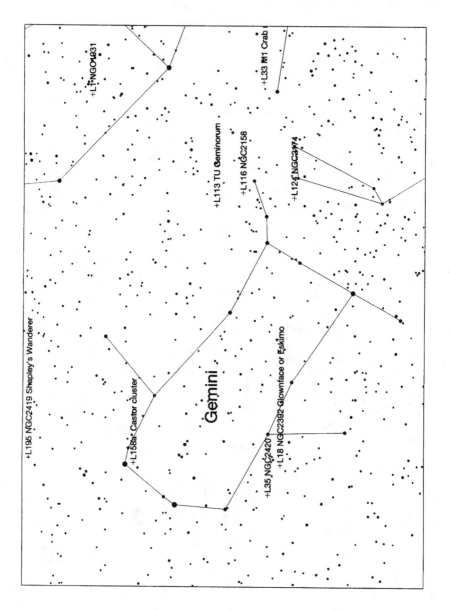

**Chart 10:
Gemini**

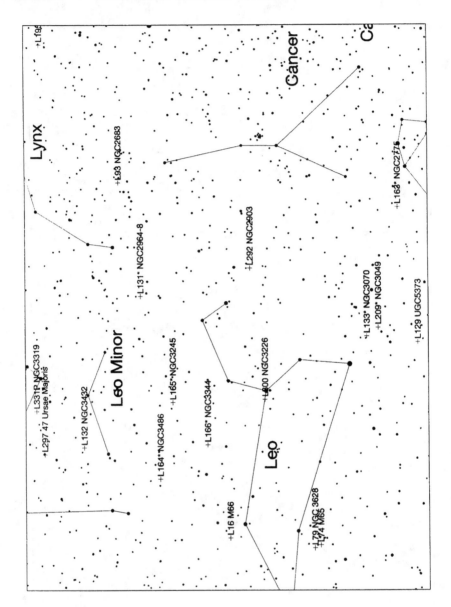

**Chart 11:
Cancer, Leo,
Leo Minor,
and Lynx**

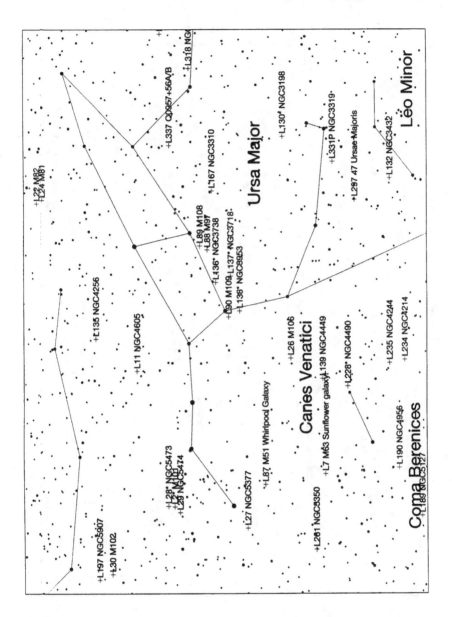

Chart 12:
Ursa Major
and Canes
Venatici

Ursa Major

Leo Minor

Canes Venatici

Coma Berenices

+L22 M82
+L24 M81

+L337 Q0957+56A/B
+L318 NG

+L167 NGC3310

+L130⁴ NGC3198

+L331P NGC3319·
+L297 47 Ursae-Majoris
+L132 NGC3432

+L89 M108
+L88 M97
+L136· NGC3738
+L137·NGC3718·
+L90 M109 +L138· NGC8953

+L135 NGC4256

+L11 NGC4605

+L26 M106

+L244 NGC4244
+L234 NGC4214

+L228· NGC4490

+L7 M63 Sunflower galaxy +L139 NGC4449

+L87 M51 Whirlpool Galaxy

+L190 NGC4956

+L28· NGC5473
+L29 NGC5474

+L27 NGC5377

+L261 NGC5350

+L197 NGC5907
+L30 M102

+L189 NGC5377

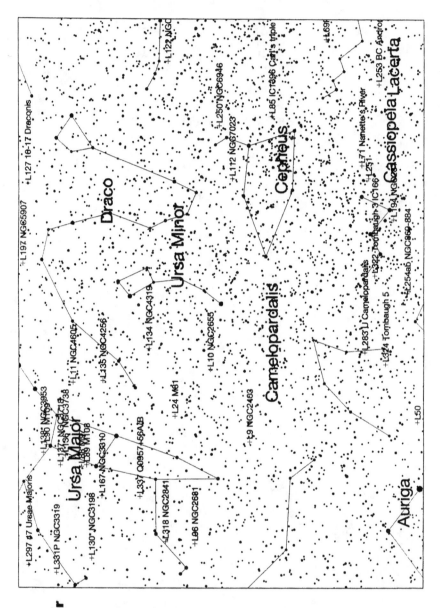

Chart 13:
North
Circumpolar

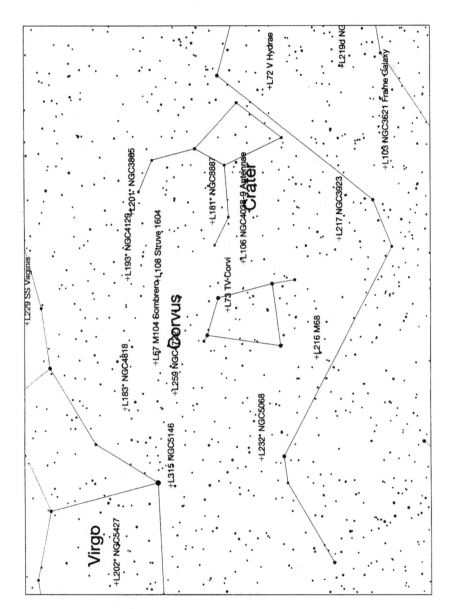

**Chart 14:
Corvus,
Crater,
southern
Virgo, and
eastern
Hydra**

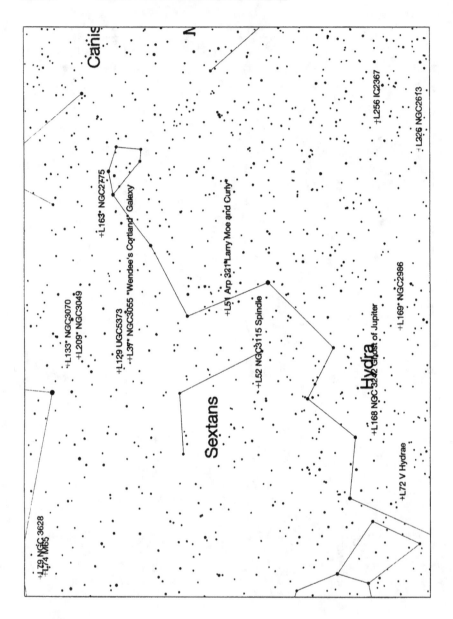

Chart 15:
Western
Hydra and
Sextans

Canis

N

+L256 IC2367

+L226 NGC2613

+L163* NGC2775

+L51 Arp 321"Larry Moe and Curly"

+L169* NGC2986

+L133* NGC3070

+L209* NGC3049

+L129 UGC5373

+L3** NGC3055 "Wendee's Cortland" Galaxy

+L52 NGC3115 Spindle

Hydra

Ghost of Jupiter

+L168 NGC 3242

Sextans

+L72 V Hydrae

+L79 NGC 3628

+L74 M65

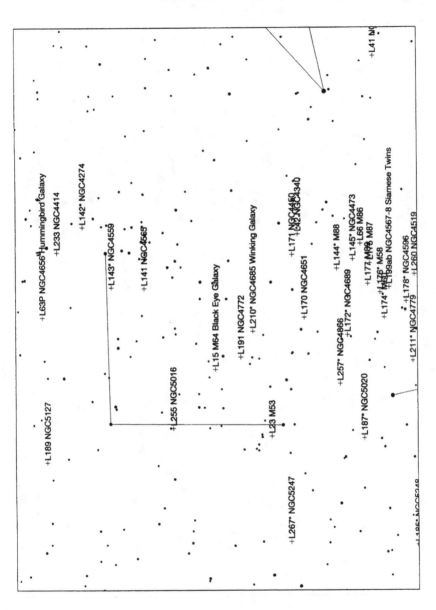

**Chart 16:
Coma
Berenices
and
northern
Virgo**

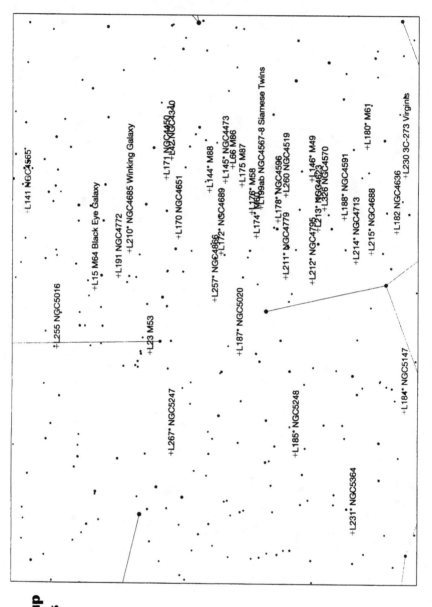

Chart 17:
Virgo group
of galaxies

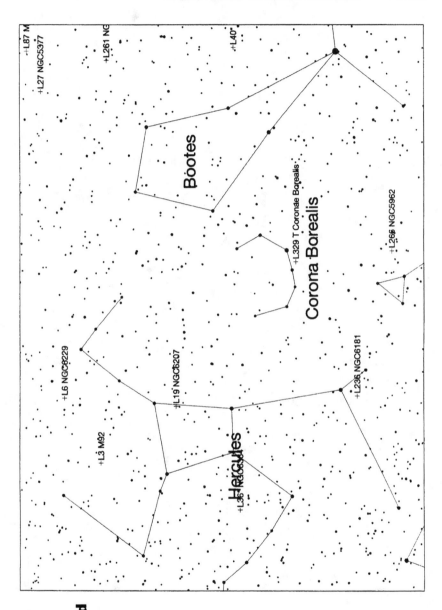

**Chart 18:
Bootes,
Corona
Borealis, and
Hercules**

Bootes

Corona Borealis

Hercules

+L87 M
+L27 NGC5377
+L261 NG
+L40°
+L329 T Coronae Borealis
+L266 NGC5962
+L6 NGC6229
+L19 NGC6207
+L236 NGC6181
+L3 M92
+L35

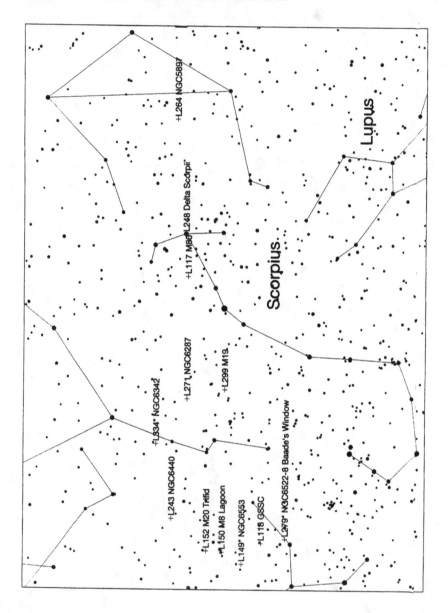

Chart 19: Libra, Scorpius, and western Sagittarius

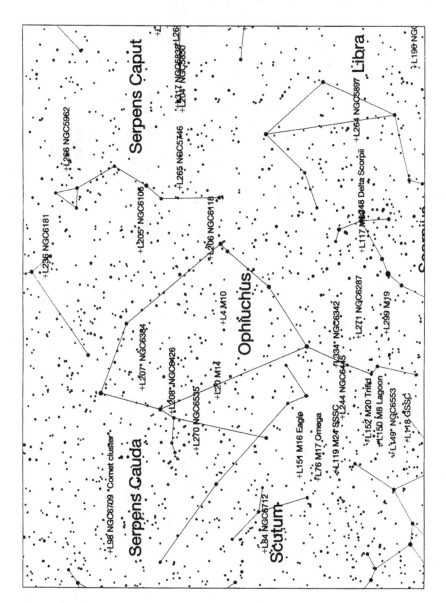

Chart 20:
Serpens
Cauda,
Ophiuchus,
Serpens
Caput

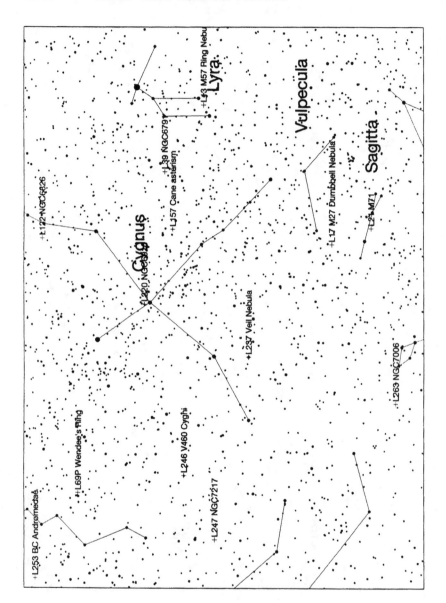

**Chart 21:
Lyra,
Cygnus,
Vulpecula,
and Sagitta**

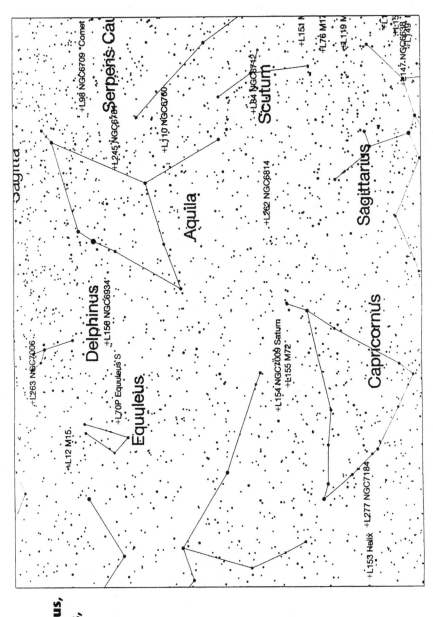

Chart 22:
Aquila,
Capricornus,
Delphinus,
and
Equuleus

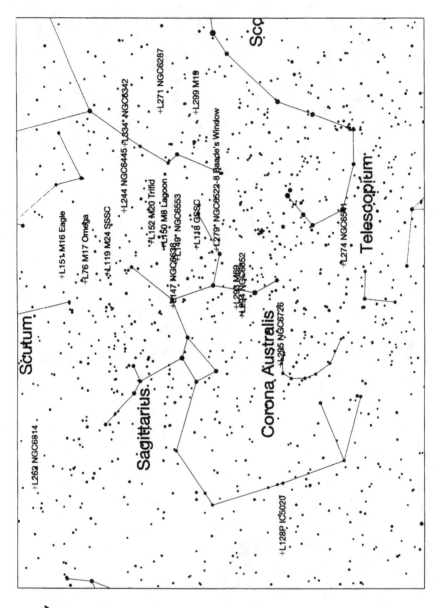

Chart 23:
Sagittarius,
Tele-
scopium,
and region

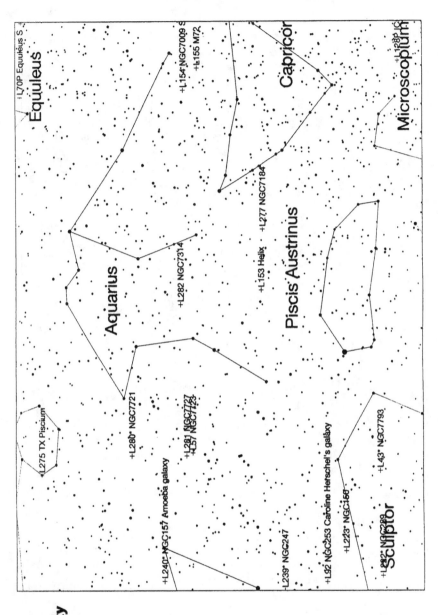

Chart 24:
Aquarius
and vicinity

Equuleus

•L70P Equuleus S

Capricor

Microscopium

•L128P PG

Aquarius

•L154* NGC7009 S

•L155 M72

•L282 NGC7314

•L277 NGC7184

•L153 Helix

Piscis Austrinus

•L275 TX Piscium

•L280* NGC7721

•L281 NGC7727

•L57 NGC7723

•L240* NGC157 Amoeba galaxy

•L239* NGC247

•L92 NGC253 Caroline Herschel's galaxy

•L223* NGC160

•L43* NGC7793

Sculptor

•L82* NGC288

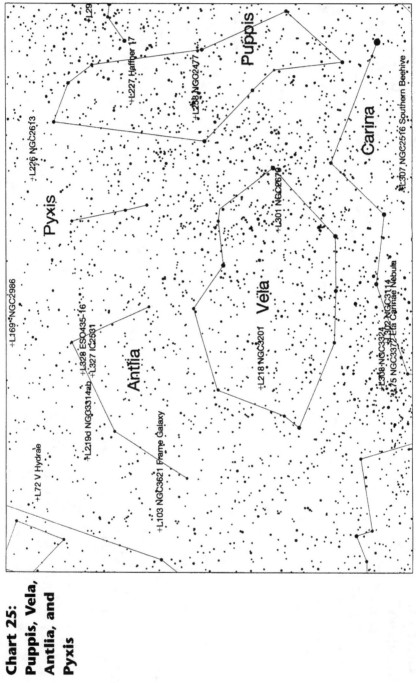

Chart 25:
Puppis, Vela,
Antlia, and
Pyxis

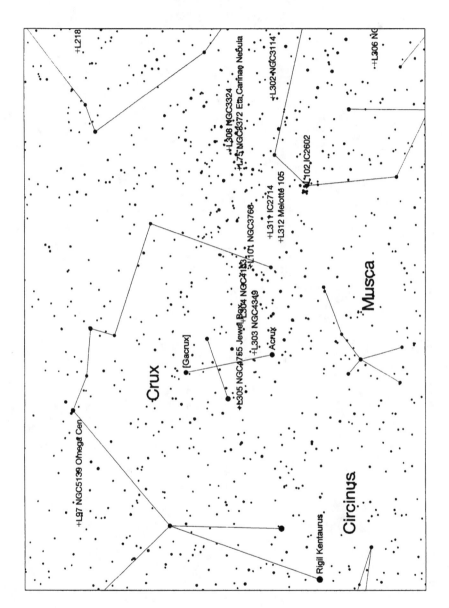

Chart 26:
Centaurus,
Crux, and
western
Carina

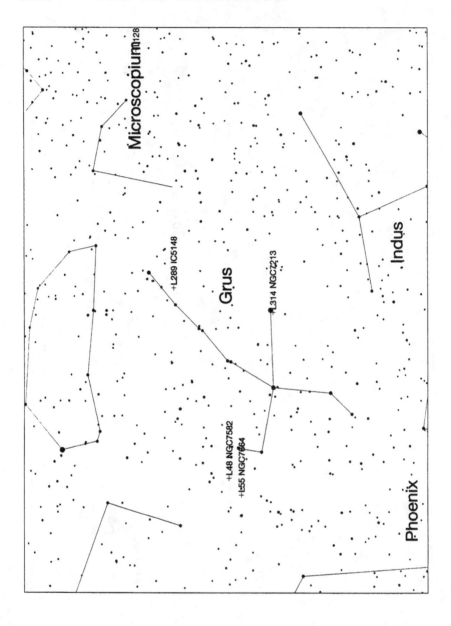

Chart 27:
Grus

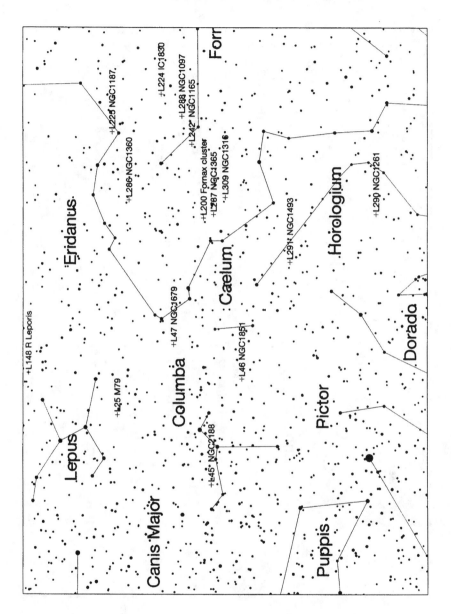

Chart 28:
Caelum,
Columba,
Pictor, and
region

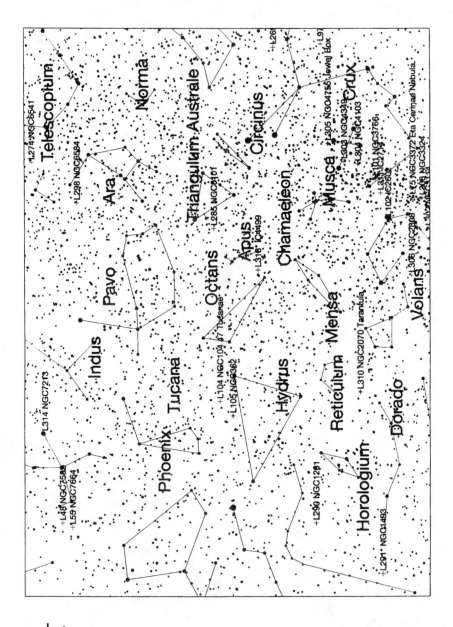

**Chart 29:
South Cir-
cumpolar**

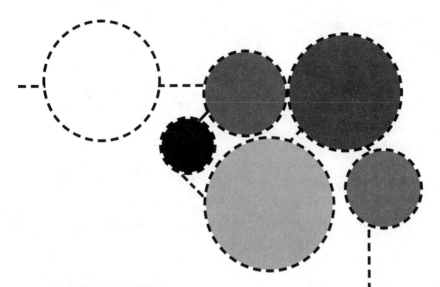

NOTES

PREFACE

1. Ken Graun, *The Next Step: Finding and Viewing Messier's Objects* (Tucson: Ken Press, 2005), p. 332.

CHAPTER I

1. CN-3 Record Book, 1965.
2. Brian Marsden, "The Comet Pair 1988e and 1988g," unpublished paper.
3. Donald K. Yeomans, *Comets: A Chronological History of Observation, Science, Myth, and Folklore* (New York: Wiley, 1991), p. 410.

CHAPTER 2

1. Leslie Peltier, *Starlight Nights* (1965; Cambridge, MA: Sky Publishing, 1999), p. 69.

CHAPTER 4

1. Brent Archinal and Steven Hynes, *Star Clusters* (Richmond, VA: Willmann-Bell, 2003), pp. 143–44.

2. Ibid., p. 143.

3. Ibid., p. 144.

4. Richard Hinckley Allen, *Star Names: Their Lore and Meaning* (1899; New York: Dover, 1963), p. 332.

5. Carolyn Gilman, "John Goodricke and His Variable Stars," *Sky & Telescope* 56, no. 5 (November 1978): 400–403.

6. Archinal and Hynes, *Star Clusters*, p. 143.

7. Arthur Preston Hankins, *Cole of Spyglass Mountain* (New York: Dodd, Mead, 1923).

8. Leslie Peltier, *Starlight Nights* (1965; Cambridge, MA: Sky Publishing, 1999), p. 193.

CHAPTER 5

1. Brent Archinal and Steven Hynes, *Star Clusters* (Richmond, VA: Willmann-Bell, 2003), p. 134.

2. Jane Houston Jones, "Lake Sonoma 2/22/03: Observing 175A-227," http://observers.org/reports/2003.02.22.7.html (accessed January 1, 2005).

3. Ibid.

4. David H. Levy, Observing Session Log, Session *6235E.

5. Stephen J. O'Meara, *Deep Sky Companions: The Caldwell Objects* (Cambridge, MA: Sky Publishing, 2002), p. 58.

CHAPTER 6

1. Stephen J. O'Meara, *Deep Sky Companions: The Messier Objects* (Cambridge, MA: Sky Publishing, 1998), p. 78.

2. *Monthly Notices of the Royal Astronomical Society* 82 (1922): 246.

3. Stephen J. O'Meara, *Deep Sky Companions: The Caldwell Objects* (Cambridge, MA: Sky Publishing, 2002), p. 182.

4. O'Meara, *Deep Sky Companions: The Messier Objects*, p. 59.

5. B. J. Bok and E. F. Reilly, "Small Dark Nebulae," *Astrophysical Journal* 105 (1947): 255.

6. See "The Horsehead Nebula: A Bok Globule in the Making?" *Sky & Telescope* 69 (1985): 12; "Bok Globules and Star Birth," *Sky & Telescope* 71 (1986): 147; and "Bart Bok Was Right," *Sky & Telescope* 81 (1991): 485–86.

7. W. Baade to B. J. Bok, March 25, 1947.

8. B. J. Bok to W. Baade, April 7, 1947.

9. W. Baade to B. J. Bok, April 21, 1947.

10. "Thousands of Suspected Globules," *Sky & Telescope* 15, no. 4 (1956): 170.

11. J. B. Sidgwick, *Introducing Astronomy* (London: Faber, 1959), p. 161.

12. C. Robert O'Dell, *The Orion Nebula: Where Stars Are Born* (Cambridge, MA: Belknap Press of Harvard University Press, 2003), p. 108.

13. David DeVorkin, *Interview with Dr. Bart J. Bok* (New York: American Institute of Physics, 1979), p. 62.

14. O'Dell, *The Orion Nebula*, p. 85.

15. James Joyce, *Ulysses* (1922; New York: Penguin, 1968), p. 619.

CHAPTER 7

1. Leslie Peltier, *Starlight Nights* (1965; Cambridge, MA: Sky Publishing, 1999), p. 231.

2. Stephen J. O'Meara, *Deep Sky Companions: The Caldwell Objects* (Cambridge, MA: Sky Publishing, 2002), p. 155.

3. Although this story appears in several sources, a particularly good one is Kenneth Glyn Jones, *Messier's Nebulae and Star Clusters*, 2nd ed. (London: Cambridge University Press, 1991), p. 365. Some of the details of Messier's relationship with de Saron come from this source as well.

CHAPTER 8

1. Bertram Schwarzchild, "Stellar Motion Very Near the Milky Way's Central Black Hole," *Physics Today* 51 (March 1998): 21.

2. Stephen J. O'Meara, *Deep Sky Companions: The Messier Objects* (Cambridge, MA: Sky Publishing, 1998), pp. 93–96.

3. An interesting historical account of this early work is in L. Kühn, *The Milky Way: The Structure and Development of Our Star System* (New York: Wiley, 1982), pp. 17–19.

4. B. J. Bok, *The Distribution of the Stars in Space* (Chicago: University of Chicago Press, 1937), p. 124.

5. Walter Baade to B. J. Bok, February 8, 1949.

6. O. Struve, "Galactic Exploration by Radio," *Sky & Telescope* 11, no. 9 (1952): 215.

7. B. J. Bok, "Radio Studies of Interstellar Hydrogen," *Sky & Telescope* 13, no. 12 (1954): 408.

8. Bart Bok, personal communication, 1979.

9. E. J. Maggio, personal communication, 1983.

CHAPTER 9

1. Stephen J. O'Meara, *Deep Sky Companions: The Caldwell Objects*, (Cambridge, MA: Sky Publishing, 2002), p. 155.

2. Ibid., p. 414.

3. Ibid., p. 416.

4. Ibid., p. 410.

5. B. J. Jones and L. G. Boyd, *The Harvard College Observatory: The First Four Directorships, 1839–1919* (Cambridge, MA: Belknap Press, 1971), p. 367.

6. Henrietta Swan Leavitt, "1,777 Variables in the Magellanic Clouds," *Annals of the Astronomical Observatory of Harvard College* 60, pt. 4 (1908): 107.

7. Harlow Shapley, *Through Rugged Ways to the Stars* (New York: Scribner's Sons, 1969), pp. 52–53.

CHAPTER 11

1. Stephen J. O'Meara, *Deep Sky Companions: The Messier Objects* (Cambridge, MA: Sky Publishing, 1998), p. 231.

2. Ibid., p. 160.

3. W. Williams, letter, *Sky & Telescope* 38 (1969): 376.

CHAPTER 12

1. I. K. Williamson, "The Messier Club," *Skyward* (August 1966): 7.

2. S. K. Vsekhsvyatskii, trans., *Physical Characteristics of Comets* (Jerusalem: Israel Program for Scientific Translations, 1964), p. 215; *Illustrated London News*, August 16, 1862, p. 179.

3. E. M. Xilouris et al., "Are Spiral Galaxies Optically Thin or Thick?" *Astronomy and Astrophysics* 344 (1999): 868–78.

4. At Tennessee's Arthur J. Dyer Observatory, Robert Hardie noted in "The Story of the Early Life of E. E. Barnard" that the university named one of its dormitories after the great comet finder, who was known not to need much sleep.

CHAPTER 13

1. This essay was adapted from my Star Trails column in *Sky & Telescope* (February 2002): 72–73.

2. Ron Cowen, "Grand Illusion," *Science News* 168, no. 4 (July 23, 2005): 62.

3. Dennis Walsh, Bob Carswell, and Ray Weymann, "09.57 + 561 A, B: Twin Quasistellar Objects or Gravitational Lens?" *Nature* 279, no. 5712 (May 31, 1979): 381.

4. Brent Archinal, message to amastro newsgroup, October 21, 2003, http://groups.yahoo.com/group/amastro/message/11127 (accessed December 1, 2004).

CHAPTER 14

1. Judith Irwin, personal communication, January 23, 2005.

GLOSSARY

Asteroid: A small object, composed of rock, orbiting the Sun, and usually irregular in shape.

Astronomical unit: The average distance between Earth and the Sun, about 93 million miles or 150 million kilometers.

Big Bang: The favored theory for the origin of the Universe.

Binary star: Two stars held together by their mutual gravity.

Black hole: An object so massive and dense that no radiation can escape from it.

Comet: A small body, usually the size of a village, made up of ice, rock, and dust. Comets orbit the Sun in highly elliptical or parabolic paths.

Dwarf star: A small star, like the Sun.

Fireball: A meteor brighter than Venus.

Galactic star cluster (or open star cluster): A group of dozens, or hundreds, of stars moving through space as a unit.

Galaxy: A monstrous collection of billions of stars, along with gas and dust. Galaxies are classified as irregular, spiral, or elliptical.

Globular star cluster: A large group of hundreds of thousands of stars, spherically shaped, and typically orbiting a galaxy.

Gravitational lens: A massive object, like a galaxy, positioned between Earth and a more distant object. As light from the distant objects passes by the lens, it is distorted. As a result, we see stretched, double, or multiple images of the distant object.

Hertzsprung-Russell diagram: A graph that plots the temperature (or color) of a star against its luminosity.

Local Group: A grouping of about thirty galaxies; the Milky Way is one of the larger members.

Local Supercluster: A gigantic collection of clusters of galaxies; the Local Group is one of the smaller members.

Main sequence: The band on the Hertzsprung-Russell diagram where most stars, including those like the Sun, are plotted.

Meteor: The term for a small meteoroid that is heated to incandescence as it enters Earth's atmosphere.

Meteorite: A piece of small interplanetary rock (meteor) that crashes into Earth's surface.

Milky Way: (1) A glowing band of milky light encircling the sky. The Milky Way is the plane on which most of the stars in our galaxy lie. (2) The galaxy in which we live.

Neutron star: The remains of a massive star that has collapsed; it consists almost entirely of neutrons.

Nova: In a binary system, a white dwarf star that brightens explosively.

Planetary nebula: A shell of gas released by an old star.

Quasar: A quasi-stellar object but far more massive than an ordinary star. Probably the active core of a distant galaxy.

Red giant: A reddish or orange old star, off the main sequence.

Supernova: An explosion produced by the collapse of a massive star, as it releases all its outer layers. At its brightest, a supernova can outshine all the several hundred billion stars of its galaxy put together.

Supernova remnant: The debris thrown off by a supernova.

Variable star: A star that changes in brightness.

White dwarf: A small, hot remnant that is left after a red giant expels its outermost layers.

GENERAL INDEX

OBJECT INDEX